THE RADIO AMATEUR'S
MICROWAVE COMMUNICATIONS
HANDBOOK

DAVE INGRAM, K4TWJ

TAB BOOKS Inc.
Blue Ridge Summit, PA 17214

Other TAB Books by the Author

No. 1120 *OSCAR: The Ham Radio Satellites*
No. 1258 *Electronics Projects for Hams, SWLs, CBers & Radio Experimenters*
No. 1259 *Secrets of Ham Radio DXing*
No. 1474 *Video Electronics Technology*

FIRST EDITION

FIRST PRINTING

Copyright © 1985 by TAB BOOKS Inc.

Printed in the United States of America

Reproduction or publication of the content in any manner, without express permission of the publisher, is prohibited. No liability is assumed with respect to the use of the information herein.

Library of Congress Cataloging in Publication Data

Ingram, Dave.
    The radio amateur's microwave communications handbook.

    Includes index.
    1. Microwave communication systems—Amateurs' manuals.
I. Title.
TK9957.I54   1985   621.38'0413   85-22184
ISBN 0-8306-0194-5
ISBN 0-8306-0594-0 (pbk.)

Contents

	Acknowledgments	v
	Introduction	vi
1	**The Amateur's Microwave Spectrum**	1
	The Early Days and Gear for Microwaves—The Microwave Spectrum—Microwaves and EME—Microwaves and the Amateur Satellite Program	
2	**Microwave Electronic Theory**	17
	Electronic Techniques for hf/vhf Ranges—Electronic Techniques for Microwaves—Klystron Operation—Magnetron Operation—Gunn Diode Theory	
3	**Popular Microwave Bands**	29
	Circuit and Antennas for the 13-cm Band—Designs for 13-cm Equipment	
4	**Communications Equipment for 1.2 GHz**	42
	23-cm Band Plan—Available Equipment—23-cm DX	
5	**Communications Equipment for 2.3 GHz**	51
	Setting Up a 2.3-GHz Amateur System—Expanding the 2.3-GHz System—QRP at 2.3 GHz—Antennas for 2.3 GHz	
6	**Communications Equipment for 10 GHz**	63
	A Beginner's Setup for 10 GHz—A Quick and Easy 10-GHz Communications Setup—A High-Quality 10-GHz Communications Setup—A Phase-Locked 10-GHz Setup for Long-Distance Communications	

7 Microwave Networking and Data Packeting   85
Computer Communications—An Expandable Network for Multimode Communications—Packet Communications

8 Power Supplies for Microwave Systems   95
Transformers—Capacitors—Regulators—A Rugged General-Purpose Power Supply—Safe-Stop Power Supply—The Pick-A-Volt Supply—Nickel-Cadmium Batteries—Natural Power Sources

9 Setting Up, Tuning, and Operating Microwave Systems   111
Characteristics of 2.3 GHz and Lower—Safety Considerations—Expansions and Refinements for Microwave Systems

10 Interfacing Microwaves With Television and Computers   119
Fast-Scan TV at 2.3 GHz—Fast-Scan TV at 10 GHz—Scan-Converting Relays—Linking Home Computers via Microwaves

11 Amateur RADAR and Intruder Alarms   128
RADAR Types—Intruder/Motion-Detector Alarms—10-GHz Mini-RADAR Concepts—10-GHz Amateur Weather RADAR

12 Microwave Exclusive: TVRO and MDS   141
The Television Broadcasting Satellites—Home Satellite-TV Reception—MDS: What It Is—Operational Concepts of MDS—MDS-Band Equipment

Appendix A   AMSAT   153

Appendix B   Phonetic Alphabet   159

Appendix C   International "Q" Signals   160

Appendix D   Great Circle Bearings (Beam Headings)   163

Appendix E   International Prefixes   168

Index   182

Acknowledgments

Creating a book of this nature isn't a simple or easy matter. I would thus like to extend sincere appreciation to the following for their assistance and information included in this book: Tom O'Hara, W6ORG; P C Electronics of Arcadia, California; Fred Stall and the gang at KLM Electronics of Morgan Hill, California; Steve and Deborah Franklin of Universal Communications, Arlington, Texas; Alf Wilson and the publishers of *ham radio* magazine in Greenville, New Hampshire; and Jim Hagan, WA4GHK, of Palm Bay, Florida.

Thanks also to Microwave Associates of Burlington, Massachusetts; Jay Rusgroove of Advanced Receiver Research; and Paul DeNapoli, WD8AHO, of ENCON, Inc., Livonia, Michigan.

Finally, a very special thanks to my XYL, Sandy, WB4OEE, for bearing (and somehow surviving) the tribulations of typing this eighth manuscript on another unique amateur-radio frontier. Here's wishing all of you the very best luck and success in your microwave pioneering.

Introduction

Somewhere on a country mountaintop an amateur microwave repeater system sits in a heated and dimly illuminated building. The large dish antennas outside the building give only a brief glimpse of the futuristic activity happening inside. The system is handling various communications, ranging from computer interlinks and amateur-television operations to multiple voice relays between various cities. An additional dish antenna is relaying signals to wideband amateur communications satellites placed in geostationary orbits at various points around the world. Many miles away, the amateurs accessing this system use small hand-held transceivers, or computer terminals; yet their operations can reach the world. Fantasyland? Indeed not; this is the shape of amateur trends that are being developed and activated at this time. Operational concepts of these systems are outlined in this book. I sincerely hope you find this information both beneficial and inspiring. Once involved in microwave pioneering, you'll surely agree this is amateur radio's ultimate frontier. As I've said many times—in previous books and in magazine columns—the Golden Age of Radio is very much alive and well. It lives in the highly specialized areas of modern communications technology.

    Involvement with amateur microwaves need not be highly technical or overly expensive. The idea, and the format of this book, is thus oriented towards enjoyment in the least expensive manner. This isn't by any criteria projected as a final word; indeed, every communications frontier is an area of continual improvement and change. The logical way to join such activity is simply secure a starting point (such as this book) and progress with evolutions. I hope the ideas herein inspire your ingenuity and creativity, and I look forward to hearing of your works.

Chapter 1

The Amateur's Microwave Spectrum

The electromagnetic spectrum of microwave allocations is one of the hottest and fastest-rising frontiers in amateur communications technology. This unique frontier offers a true kaleidoscope of unlimited challenges and opportunities for today's innovative amateurs. Although a relatively uncharted area until recent times, today's microwave spectrum is gaining a widespread popularity and rapidly increasing acceptance. This trend shows no signs of waning; indeed, microwave communications are destined to mark the path of future developments in amateur communications. These communications will include all modes, from data packeting and multichannel television relays to multichannel voice links of FM, SSB, and computer interlinks. While the line-of-sight propagation associated with microwave communications would seem to restrict its capabilities, such is not necessarily the case. This situation has been commercially exemplified in such arrangements as long-distance telephone microwave links, television microwave networks, etc. These systems provide broadband cross-country and intercontinental linking. Transcontinental linking has been accomplished by geostationary communications satellites. Amateur radio is destined to progress in a similar manner; furthermore, amateur satellites capable of providing these interconnect functions are being developed at this time. The future of amateur radio looks quite promising and very exciting, and microwave communications will play a major role in its developments.

The h-f band operator of today might ponder the logic of using microwave communications. Why switch from the populated rf areas to a seemingly vast, empty, range of extremely-high-frequency spectrum, when few amateurs operate that range? One reason is that the line-of-sight propagation of microwaves affords reliable and predictable communications, independent of solar or weather conditions. Extended communication ranges are possible using one, two or more microwave repeaters. Additionally, the wide bandwidth associated with such repeaters allows multiple communications to be simultaneously conducted.

The following example may further clarify this situation: Assume two amateurs living in metropolitan areas separated by one (or two) mountains. They desire to set up a fast-scan-television repeater station. Although an in band 70-cm (420 MHz) system could be used, it would require expensive filters and duplexers for effective operation, and that operation would carry only one transmitting signal at a time. A crossband fast-scan repeater operating with an input on 70 cm and an output on 23 cm (1240 MHz) or 13 cm (2300 MHz) would alleviate the problems and costs of special filters and duplexers. However, its operation would still be confined to only one transmitting signal at a time. Thinking ahead, the two amateurs would set up a relatively inexpensive 10 cm (2300 MHz) or 3 cm (10,000 MHz, or 10 GHz) "bare bones" repeater station for relaying their signals across the mountainous area. At any later time, other amateurs could join the activity simply by adding the appropriate microwave "front ends" to their setup. An additional microwave link could then be added at one operator's location for further feeding the signals to other interested amateurs. Each new addition to the network would carry its own weight in equipment support/finance, causing the system to grow and expand precisely in the direction of most interest. The original two network-instigating amateurs are now part of a multioperator system.

Further, let's assume several amateur-radio computer enthusiasts, plus some amateur RTTY (radio teletype) operators, and a number of voice-only operators desire to join the network. The vast bandwidth capability of this system stands ready to accommodate the new group of amateur operators: only minor alterations in power levels and antenna configurations are necessary.

The network continues to grow until several communities and cities are linked in a totally reliable and predictable manner. An amateur satellite uplink/downlink is added to the network, along with electronic-mailbox and intelligent-voting systems, plus emergency/priority interrupts for special requirements. The network ultimately spans coast to coast and continent to continent, conveying many forms of amateur-radio activity. Each new area would be responsible for its own expenditures, and thus the system carries its own weight. The original instigators, plus many fellow operators, now enjoy multimode communication from small, personal, transceivers that access the network via simple 2-meter, 70-cm, or newly introduced 13-cm units.

Science fiction? Hardly. A vision into the near future? Surely. Realizing the many beneficial aspects of microwaves, only one of which has been exemplified here, we can truly calculate that amateur operations during this and subsequent decades will flourish through utilization of all available assets—and the microwave spectrum is one of these prime assets. A simplified example of the previous discussion is shown in Fig. 1-1.

Moving in a slightly different direction, let's now consider a more personal application for which microwaves could again prove useful. An individual microwave link can be used for remote high-frequency receiving setups. Several wideband converters, for example, can be connected to respective antennas and used for reception of all hf bands. The resultant wideband spectrum may then be microwave relayed to an amateur's home location or transmitter site. Following retrieval of the h-f spectrum from the microwave receiver's output, conventional signal processing can be utilized for producing a truly optimum DXing setup. The signal diversity creates unique capabilities which thus allow a station to perform in a definite "top-gun" manner. See Fig. 1-2.

THE EARLY DAYS AND GEAR FOR MICROWAVES

Although a little known fact, experiments in the microwave spectrum date to the very early days of radio pioneering. A number of Heinrich Hertz's early experiments with "Hertzian waves" during the late 1800s were at wavelengths which translate to frequencies of between 400 and 800 MHz. Guglielmo Marconi's early European experiments in radio utilized simple spark-gap equipment with small coils; the accompanying receiver also used basic "hooks" of wire. Translating the physical dimensions of this primitive gear to its corresponding wavelength and frequency yields an rf spectrum of approximately 1.5 to 3.0 GHz. Microwave communications have, indeed, been with us since the early days of radio activity.

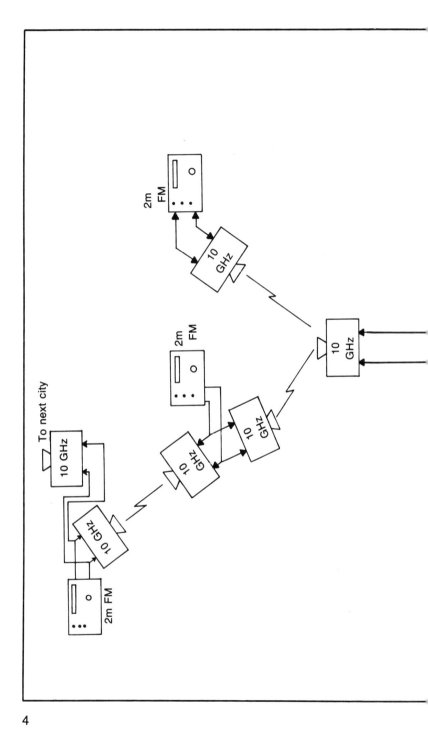

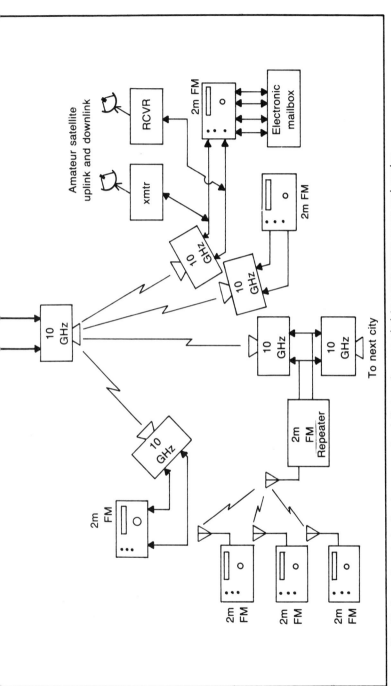

Fig. 1-1. Simplified overview of a basic microwave network that can be expanded to cover many areas and modes.

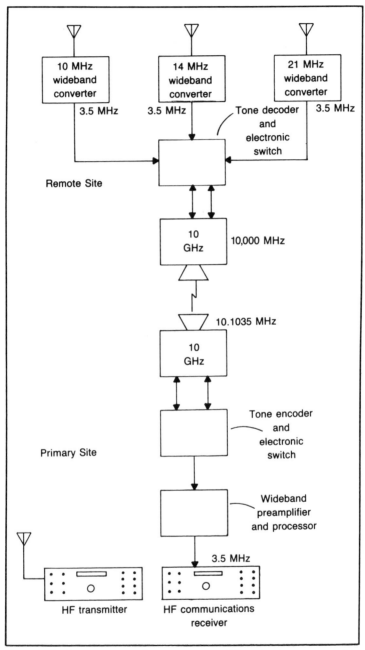

Fig. 1-2. Basic arrangement for a remote receiving site linked by 10-GHz microwave equipment.

Continuing toward our present period of time, we find a somewhat crude version of the magnetron tube developed during the mid-1920s. This unique tube used a strong magnetic field, created by large magnets surrounding the device, to deflect electrons from their natural path and thus establish oscillation in the microwave range. Because specific technology wasn't yet available for putting the device to use, however, the magnetron laid (basically) dormant for several additional years.

The European continent was also reflecting significant pioneering efforts in the microwave spectrum. A 2-GHz link was operated across the English Channel during the mid 1930s. During the 1940s, the cavity magnetron was devised and placed into use with the first RADAR (RAdio Detecting And Ranging) systems.

Ensuing evolutions during subsequent decades produced the klystron, reflex klystron, the traveling-wave tube, the Gunn diode, and the recent GaAsFET transistors. The difficulties in developing these microwave devices revolve primarily around electron transit time for each cycle of wave propagation. Stated in the simplest of terms, electrons leaving the cathode of a tube (and traveling toward that tube's plate), must transit a path shorter than one-half wavelength. This situation is not of consequence in low-frequency devices; however, an alteration of design is required for microwave operations. The lighthouse tube (by General Electric), and acorn tubes were introduced to fulfill this need. By directing electron flow in more direct patterns while reducing stray and interelectrode capacitance, these devices allowed microwave operations at frequency ranges that were previously not feasible. As knowledge expanded, higher and higher frequencies became practical. The restrictions of stray capacitances and transit times were overcome, and "tuned circuits," such as they are for these extremely-high frequencies, were incorporated directly into the new devices.

THE MICROWAVE SPECTRUM

The frequencies comprising the microwave spectrum extend from approximately 1,000 megahertz, or 1 gigahertz, to approximately 50,000 megahertz, or 50 gigahertz. The upper end of this range is somewhat undefined, and indeed unpioneered, when visualized in respect to general amateur applications. A list of amateur frequencies available is shown in Fig. 1-3. While the 144, 220, and 432 MHz allocations are not microwave frequencies, they are included here as a reference to known and established amateur

2m - 144 - 148 MHz		.144 - .148 GHz
1 1/4 m - 220 - 225 MHz	Reference only	.22 - .24 GHz
70 cm - 420 - 450 MHz		.42 - .45 GHz
46 cm - 860 - 890 MHz		.86 - .89 GHz
23 cm - 1,240 - 1,300 MHz		1.24 - 1.3 GHz

Commercial weather satellite range
1.690 - 1.691 GHz
13 cm - 2,300 - 2,450 MHz   2.3 - 2.45 GHz

MDS band
2,100 - 2,200 MHz   2.1 - 2.2 GHz
10 cm - 3,300 - 3,500 MHz   3.3 - 3.5 GHz

Satellite TV band
3,700 - 4,200 MHz   3.7 - 4.2 GHz
5 cm - 5,650 - 5,925 MHz   5.65 - 5.925 GHz
3 cm - 10,000 - 10,500 MHz   10.0 - 10.5 GHz

X band
10,500 - 10,600 MHz   10.5 - 10.6 GHz

15 mm
24,000 - 24,500 MHz   24.0 - 24.5 GHz

K band
48,000 - 48,500 MHz   48.0 - 48.5 GHz

Fig. 1-3. Frequency allocations in the microwave spectrum.

areas. Likewise, the MDS and satellite TV bands (2.1 and 4 GHz) are shown as a means of familiarizing the amateur with the microwave spectrum.

The Low End

Almost every amateur is familiar with the 144-MHz (2-meter) amateur band. FM, SSB, and amateur-satellite communications are used rather extensively in this range throughout the United States and most of the world. As the 2-meter band filled with amateur activity, operations expanded to 220 MHz. As a number of FM repeaters became operational in this spectrum, activity once again expanded to include the 440-MHz (70-cm) amateur band. The 70-cm band is primarily used for FM, amateur fast-scan television, and OSCAR (Orbital Satellite Carrying Amateur Radio) amateur communications.

860 MHz

Slightly higher in frequency, the next amateur band is 860 to 890 MHz. This allocation was acquired as this book was being written, thus its applications and future in amateur radio are unknown at the present time. This band is expected to become an amateur fast-scan-TV/OSCAR-satellite range. Its proximity to the upper end of uhf television channels is particularly appealing for public-service applications during emergencies, or for public-relations use.

23 cm

The next amateur band is 23 cm, or 1240 to 1300 MHz. It should also be mentioned at this point that 1,000 MHz is equal to 1 gigahertz, or GHz. The 23-cm band may thus be referred to as 1.24 to 1.3 GHz, if desired. The 23-cm band is becoming quite popular in many areas of the United States and Japan. Numerous amateur fast-scan-TV repeaters operate near the 1265 MHz range, and Phase-IV OSCAR satellites are slated to use the lower portion of this band for uplink signals. Equipment for 23-cm operation can be relatively inexpensive if the amateur shops carefully and plans his moves. Inexpensive varactor-tripler circuits for translating a 432-MHz signal to 1296 MHz may be constructed with minimum effort, and the results are quite gratifying. Receiving downconverter "front ends" for 23 cm are available in kit form, or preassembled from several sources listed in monthly amateur magazines. Such converters usually feature high-gain, low-noise, rf sections, and relatively low purchase costs. A substantial amount of 23-cm equipment is slated to become available for amateur use in the near future, thus activity on this band is destined to significantly increase. The long-distance communication record on 23 cm stands at 1,000 miles—a feat accomplished by using temperature-inversion and signal-ducting propagation.

MDS and Satellites

Situated between the amateur 23 cm and 13 cm bands are two particularly interesting commercial services. The weather satellite band used for studying cloud formations from approximately 20,000 miles above earth employs 1691 MHz while the public carrier service of MDS (acronym for Multipoint Distribution System) employs the range of 2100 to 2150 MHz. Although reception of weather satellites has previously appealed primarily to commercial services, numerous amateurs are realizing the advantages of this capability, and are constructing their own receiving systems. Several inexpensive receiving kits have been recently introduced for satellite reception. The MDS band may best be recognized by its recently dubbed nickname of "microwave TV braodcasting." Carrying restricted-type viewing similar to cable-TV programming, microwave-TV systems operating in the 2.1 GHz range are springing up across the nation. Reception of these pay-TV signals may be accomplished through the use of relatively inexpensive 2.1 GHz downconverters. Additional information concerning this commercial activity is presented later in this book. The United States space shuttles also use the 2.2-to 2.4-GHz range during flights. Numerous educational-television services also frequent this spectrum for point-to-point relays.

13 cm

The 13-cm amateur band holds particular appeal for future amateur activities. Its proximity to the MDS band permits use of inexpensive 2-GHz downconverter receiving systems and 2.3 GHz transmitting gear in a very cost-effective manner. A group of amateurs in a given area can actually become operational on 2.3-GHz for a lower expenditure than on almost any other amateur band. Direct communications on 2.3 GHz typically range from 20 to 60 miles, depending on terrain and the antenna systems employed. This spectrum is especially attractive for such wideband signals as multichannel fast-scan TV, multiplexed data links, computer interlinks, etc. A number of 2-meter repeaters could also be linked via 2.3 GHz, and the line-of-sight propagation would permit peaceful coexistence of several of these services in any particular metropolitan area.

5 and 10 cm

The 10 cm and 5 cm amateur bands have received miniscule interest during the past, primarily due to the lack of effective gear capable of operation in this range. The recent escalation of interest in satellite-TV terminals capable of operating in the 3.7- to 4.2-GHz range, however, shows great promise in ratifying that situation. Since many telephone companies utilize frequencies between 5 and 10 cm for broadband relays of multiple voice links, evolutions may also provide a surplus of modifiable gear for radio amateurs.

3 cm

The 3-cm (10-GHz) amateur band is gaining popularity at a very creditable rate. The primary equipment used for these 10-GHz activities is the Gunnplexer. The Gunnplexer has a Gunn diode located in its 10-GHz cavity, which is directly mated with its waveguide and horn-antenna system. The complete 10-GHz unit functions as a "front end" for a lower frequency unit that acts as an i-f stage. A small portion of the transmitted signal from each Gunnplexer is used as the receiver's local oscillator. A further clarification of this technique is shown in Fig. 1-4. The two communicating Gunnplexers are frequency separated by the amount of the desired i-f, which is 146 MHz in this example. Both Gunnplexer transmitters remain on continuously, thus providing a local oscillator for mixing with the 10-GHz signal from the other unit. The ultimate result is a 146-MHz signal appearing at the i-f port of each Gunnplexer. These 3-cm communications systems have proven their abilities over paths of 100 miles (160 km), and several European amateurs have communicated over 500 km (310 miles) on 10 GHz. An attractive plaque, sponsored by Microwave Associates of Massachusetts, awaits the first 3-cm pioneers to break the 1000-km (621 mile) range on this unique band. Gunnplexer communication networks are ideally suited for data communication links and multichannel TV relays, and as such could truly mark the direction for future developments in amateur communications.

Higher Bands

The 15-mm and higher amateur microwave bands represent

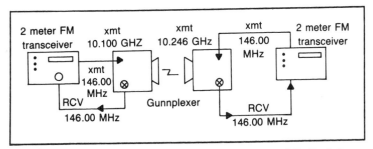

Fig. 1-4. A basic Gunnplexer communications system for 10 GHz. Each Gunnplexer oscillator provides energy for transmitted signal and couples a small amount of that energy into a mixer for heterodyning the received signal down to an i-f range. The two transmitter signals are separated by the frequency of the chosen i-f.

Fig. 1-5. Author Dave Ingram, K4TWJ, makes preliminary focal-point adjustments in a 10-GHz Gunnplexer and 3.5-foot dish antenna to be used in a microwave link. The system is capable of relaying amateur high-frequency band signals or amateur television (ATV) signals.

truly challenging and unpioneered frontiers in communications. Until recent times, the prime drawback to amateur operations in this range has been a lack of available gear, parts, and technical information. Again, Microwave Associates of Burlington, Massachusetts, has recognized this situation and provided a means of operation. Special Gunnplexers for 24 GHz and (upon special order) 48 GHz are available for less than the cost of many 2-meter transceivers. This inspiring challenge can open new doors for amateurs, and firmly establish those involved as pioneers in microwave history. What else could one ask? Yes, today's Golden Age of Radio is alive and well—particularly in the unpioneered regions of microwave communications! See Fig. 1-5.

MICROWAVES AND EME

The microwave range has, for many years, been synonymously related to amateur moonbounce activities. Centering on the 70-cm, 23-cm and 13-cm bands, amateurs have often successfully communicated over this Earth-Moon-Earth path. The parameters associated with moonbounce are many: they include considerations of atmospheric losses, faraday rotation, moon-encountered losses, galactic noise interference, etc. A general outline of these parameters is illustrated in Fig. 1-6.

The Earth-Moon-Earth distance varies between 225,000 miles (perigee) and 250,000 miles (apogee), producing fluctuations of up to 2 dB of reflected signals—a difference between communicating and not communicating via this difficult path. The EME signal is also masked by a variety of noises and requires top-notch earth-station setups plus high-gain antennas and high transmitted power levels for ensured success. The minimal acceptable rf-output power is 400 watts, and the minimal antenna-gain figure is 20 dB. These parameters do not allow any leeway for additional signal fades or noise, thus one can logically surmise that EME communications reflect extreme challenges for only the stout hearted!

The full aspects of EME communications are beyond the scope of this book, thus the reader is referred to more specialized works

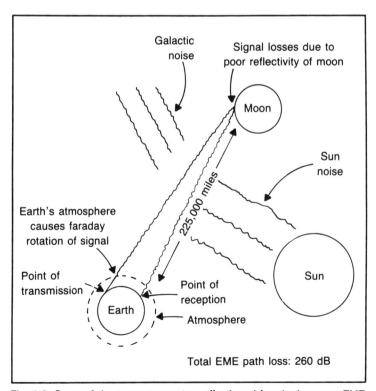

Fig. 1-6. Some of the many parameters affecting uhf and microwave EME signals.

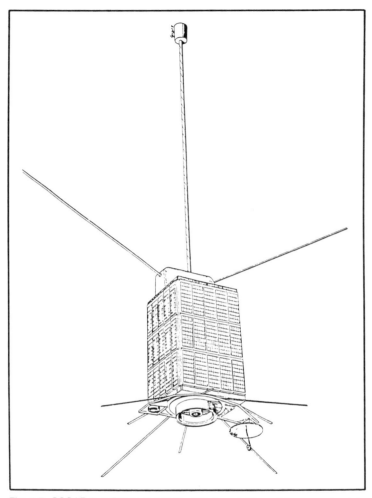

Fig. 1-7. OSCAR 8, a Phase-II Amateur Radio satellite, orbits approximately 800 miles above the Earth, where it relays 70-cm, 2-meter, and 10-meter signals. Future (Phase-III) spacecraft will use 432, 1260, and 10,000 MHz to provide hemisphere-wide communications capability.

in this particular area. Rest assured that additional information and equipment for EME operations will be a natural part of tomorrow's innovations.

MICROWAVES AND THE AMATEUR SATELLITE PROGRAM

The OSCAR satellite program utilizes several amateur microwave bands, and future projections call for yet more use of these bands. OSCAR 8, for example, produced a mode-J output on 70 cm that could easily be received by basic amateur setups. The OSCAR 9 satellite includes beacon transmitters operating in the 13-cm and 3-cm bands, which again reflects the wave of future events. OSCAR Phase III satellites are projected to afford communication capabilities in the 23-cm, 13-cm, and 3-cm bands, thus our amateur microwave spectrum may become quite popular and commonplace during the mid 1980s. See Fig. 1-7.

The microwave spectrum, with its reliable line-of-sight propagation, is particularly appealing for future geostationary (Phase III) OSCAR satellites. Relatively large dish antennas can be directed at these satellites, resulting in very dependable communications. Through the use of earth-based microwave OSCAR links, one or two spacecraft may be interlinked for near global communications. Future OSCAR satellites are destined to be recognized as prime users of amateur microwave frequency allocations.

The microwave spectrum in its entirety promises to be a major factor in future amateur-radio pioneering. The vast bandwidth

Fig. 1-8. A view of the future of Amateur Radio communications? A 10-GHz Gunnplexer and 2-meter hand-held transceivers combine to expand the horizons.

allocations, combined with computer communications and other advanced technology forms, will permit this range to be used in a heretofore unrealized manner. Dependable and reliable amateur communications with distant lands will be provided by long range OSCAR satellites, while cross-country microwave networks will provide nationwide signal linking.

Hand-help FM transceivers will also gain "seven-league boots" through microwave links and FM-to-SSB converters situated at OSCAR satellite uplink points. Also, EME systems may use moon-based microwave repeaters. Amateur pioneering efforts, however, will not cease; a creditable rise of interest in radio astronomy will serve as proof of that situation.

The following chapters of this book describe, in easy-to-understand form, the exciting world of amateur microwave operations. Separate discussions of the history of microwaves, getting started in microwaves, and detailed information on equipment and operations on various bands is included. This works is thus a guide for microwave newcomers. Here's your invitation and join the excitement of this challenging amateur frontier. Come on along and get in on the action! See Fig. 1-8.

Chapter 2

Microwave Electronic Theory

While the operational concepts associated with microwave technology are similar to their lower-frequency counterparts, this situation may seem unclear to the hf-laden amateur. Low-frequency circuits comprise physically apparent coils and capacitors of obvious dimensions. The related values for microwave-frequency applications, however, are substantially less and are usually built in to circuit layout/design rather than being interconnected by wires. This means that active devices for these frequencies will be located precisely at their associated tuned circuits (or vice versa). The changes necessary for circuit layout and design (microwave opposed to hf) is not an abrupt change, however, they evolve according to the particular frequency range(s). Stated another way, circuit designs for 220 MHz are similar to designs for 14 MHz except for the physical and electrical size of components. Circuit designs for 2300 MHz are similar to those for 145 MHz, except that coils in tuned circuits are replaced by strip lines. Likewise, 10-GHz systems are similar to 2300-MHz systems except that complete stages must be integrated directly into a cavity assembly.

When the length of a wave at microwave frequencies is considered, we realize why specific design parameters are applied. If, for example, a wavelength is only 3 cm, conventional wiring techniques would obviously kill any and all signals merely in stray capacitance and inductance (the equivalent to assembling an audio amplifier circuit within a 4 to 5 mile chassis area). Because of a number of effects, most microwave circuits, particularly those employed for amateur use, are relatively low in efficiency (typically 30 to 35 percent). Among the causes of this low efficiency are grid losses in oscillator stages, skin effect in equivalent tuned circuits (skin effect is the tendency for electrons to flow only on outer areas of conductors), etc. These will be detailed later in this chapter.

Considering the previously described aspects of microwave communications, one may thus logically surmise the majority of operation in this range could be truly categorized as a QRP and designer's haven. The challenges of designing, constructing, and using equipment in this range is, indeed, a unique experience for today's communications pioneers.

ELECTRONIC TECHNIQUES FOR hf/vhf RANGES

One of the most logical aids to understanding microwave techniques is through a review of similar hf and vhf techniques, and their subsequent relation to microwave concepts. The reasoning of this situation is quite simple; electronic operations are technically related for all frequencies, with modifications categorized according to wavelengths.

A self-excited oscillator for use on either hf or vhf requires, in addition to tuned circuits, a means of sustaining oscillation through positive (regenerative) feedback. Oscillators such as the conventional Armstrong, Hartley, etc, acquire a feedback signal directly from their associated tuned circuits, whereas oscillator circuits such as tuned grid tuned plate acquire their feedback signal from interelectrode capacitance of the tube or transistor. Since that device's output signal is fed back to its input in phase, the signal amplitude increases to provide a high output level and high efficiency. In order to sustain oscillation, two criteria must be fulfilled: an acceptable amount of interelectrode/stray capacitance must be available for providing oscillation, and the tuned circuits must exhibit resonance at the desired frequency of oscillation. Should interelectrode capacitance prove too low to provide oscillation, either slightly larger amounts of capacitance or slight changes of input/output tank circuits are usually necessary. The concept of arranging an output tank circuit near an input circuit has proven its ability to create oscillation (whether or not desired). The schematic diagram of a typical TGTP oscillator for hf/vhf is shown in Fig. 2-1.

High-frequency amplifier circuits are similar to those of oscillators, except that interelectrode capacitance is minimized and

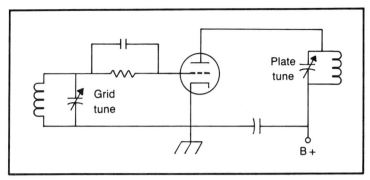

Fig. 2-1. Tuned-grid, tuned-plate oscillator for use on the high-frequency bands. Interelectrode capacitance provides feedback signal coupling to sustain oscillation.

input/output circuits are separated to prevent positive feedback. Indeed, small amounts of negative feedback are often utilized in amplifiers to prevent oscillation and improve output signal quality while assuring stable operation. These circuit requirements are usually fulfilled by such simple measures as placing all input-associated circuitry below chassis and all output-associated circuitry above chassis.

The interelectrode capacitance of an amplifying device (tube, transistor, etc.) plays a significant role in its operation. This effect is usually negligible at audio frequencies and may be ignored. For example, 50 pF could be considered of minor consequence in audio stages, but it would create problems at hf or vhf frequencies and would be considered an exorbitant value for frequencies above 1 GHz. This large capacitance could introduce positive feedback and create intolerable oscillations or it could bypass all signals to ground. The amplifier would indeed be rendered useless. See Fig. 2-2. It should be apparent from the past discussion that amplifier designs for vhf are far more critical than their hf counterparts. Significantly high output power levels may be achieved on hf as compared to vhf, because larger active devices (tubes, transitors, etc.) with consequent higher power ratings may be utilized. Vhf circuits however, require devices which exhibit lower total stray capacitance. These basic facts serve to illustrate the prime reasons why high power levels at microwave frequencies are particularly difficult to achieve.

Mixer circuits for hf and vhf ranges are, generally speaking, conventional in design. A local oscillator signal and an incoming rf signal may simply be wired to input elements of an active mixer;

the resultant output signal (sum, difference, and two original frequencies) will thus be produced at the device's output. A typical example of this arrangement is shown in Fig. 2-3. Notice that the local oscillator signal is directed to the device's emitter while the incoming rf signal is directed to the base. The signals combine in a non-linear fashion, producing the resultant sum/difference output at the collector. The interesting point of this circuit is its simplicity in design without undue concern for stray capacitance.

ELECTRONIC TECHNIQUES FOR MICROWAVES

The design and layout of oscillator circuits for microwave operations utilize extremely small values of inductance and capacitance. A tuned-grid tuned-plate oscillator for 432 MHz, for example, would typically employ a single hairpin loop for tuned circuits; the loop's stray capacitance in combination with its inductance creates a resonant tank circuit. As the operating frequency is increased, the high-inductance hairpin is replaced by a single piece of wire or strip of etched circuit-board line. A circuit of this nature is shown in Fig.

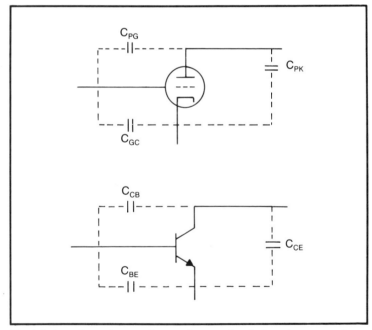

Fig. 2-2. Interelectrode capacitance of any active device plays a significant role in its operation. Such capacitance is illustrated here by the dotted-line capacitors.

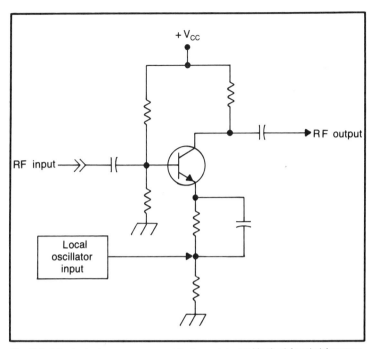

Fig. 2-3. A typical mixer arrangement that may be used in the hf and vhf ranges. The local-oscillator signal is coupled through a capacitor to the emitter, and the incoming signal is fed to the base. The intermediate-frequency (i-f) output is taken from the collector.

2-4. Note that the strip line length is directly determined the circuit's fundamental frequency of operation. As that frequency increases, strip line lengths naturally become shorter. As frequencies again move higher and into the microwave spectrum, strip lengths become critical, and active circuit components must be directly integrated in their associated tank circuits. This concept of placing active components directly into associated cavities is usually employed at frequencies of 3 GHz and higher. A microwave cavity functions as a tuned circuit because it exhibits both inductance and capacitance. The cavity's inner circular area provides inductance while the spacing between cavity top and bottom determines its capacitance. Overall physical dimensions of the cavity reflect its resonant frequency. The consideration of providing amplifier rather than oscillator action at frequencies above 3 GHz is sometimes critical; any stray capacitance/inductance may easily shunt signals to ground. Careful design with direct component location mounting is thus mandatory.

Achieving significant amounts of amplification at microwave frequencies is relatively difficult. In addition to the previously mentioned stray capacitances, plus associated skin effects, device element dimensions also govern signal handling abilities. Small transistor barrier regions, for example, limit power levels to milliwatt range; devices providing additional power handling capability thus cost more than normal amounts invested by amateurs. State-of-the-art designs use input coupling capacitors placed directly at the device. Likewise, output load and coupling capacitors are located directly at the device output points, and components are placed flat on the board and leads cut to absolute minimum length. For frequencies of 3 GHz and higher, circuits usually employ chip capacitors rather than conventional disc capacitors. The values of these chips are similar to their lower-frequency counterparts—0.01 to 0.0001 μF typical. the chip capacitors, however, are leadless and exhibit almost zero lead inductance.

Because of the physical layout of microwave mixers, these circuits appear almost mechanical in nature. Single wires or short pc board strips serve to couple signals; their physical location is usually quite critical. A wire placed near another wire may form a coulin

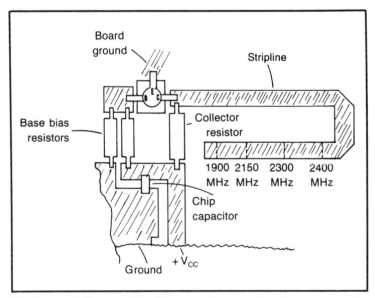

Fig. 2-4. A strip-line tuned circuit is shown in this basic 2-GHz oscillator. The length and width of the line connected to the collector determines the operating frequency of the circuit.

circuit of relatively high efficiency. Moral: mechanical and lead rigidity is a prime concern in microwave circuits. Once this balance is achieved, it must not be upset.

Because conventional active mixing devices (tubes, transistors, etc.) exhibit high element noises, they are useless at microwave frequencies. Diode mixers are thus employed. Although these diodes are also noisy, the resultant noise figures are usually acceptable. Signal mixing with a single diode is accomplished by directing both a local oscillator signal and an incoming signal to the device. The resultant i-f signal is then coupled from the diode through a frequency-selective circuit. This concept isn't new or unusual. It has been employed for years in many circuits. One example is the video detector in a television set. This diode beats, or hetrodynes, video and sound carrier frequencies to produce a resultant sound i-f center frequency. Because rf amplification at microwave frequencies is quite difficult and relatively expensive, downconverting mixer setups are very popular. This concept involves downconverting a full spectrum (noise and all) to a lower range where it may be processed in a more conventional and effective manner.

The intermediate frequencies used for microwave communications are usually tailored according to specific systems criteria. A single audio-channel link at 10 GHz might use 28.5, 29.6, 146.00, or 108 mHz as an i-f, a 10-GHz amateur video link would work well with an i-f of TV channel 2 (54 to 60 MHz) or channel 3.

KLYSTRON OPERATION

Although seldom encountered in conventional amateur setups, the klystron tube is an interesting microwave device capable of operation as an amplifier, oscillator, or mixer. The usual operating range of klystrons is from 800 MHz to 30 GHz, and their rf power levels range from milliwatts to several kilowatts. The klystron is, essentially, a complete unit within itself; input and output tuned cavities are included within the tubes construction. The average amateur may have difficulty locating klystron tubes (particularly those capable of mechanical returning for amateur bands). However, these devices occasionally appear in military surplus markets. A major consideration when acquiring a klystron involves also obtaining information and schematics necessary for operation of the device.

The klystron tube proper consists of an electron-emitting heater

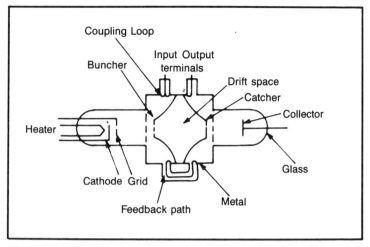

Fig. 2-5. An outline of a klystron tube. The buncher and catcher grids are connected to resonant cavities that are excited by the electrons passing through the grids. The size of the cavities and the distance between the grids determine the operating frequency of the device.

and cathode, accelerating grid, buncher grid, buncher grids, catcher grids, and a collector plate. The buncher grids are placed at each end of an input cavity, and the catcher grids are placed at each end of an output cavity. A drift area is situated between input and output cavities. An outline of this tube is shown in Fig. 2-5.

The klystron's resonsant cavities function as tuned circuits, their excitation being provided by and the electrons flowing through the grids. As electrons leave the accelerating grid and move toward the collector plate, they first encounter the two buncher grids. Each half cycle of energy thus sets up oscillations between the catcher grids, assisting or impeding electron flow through the drift area. As the accelerated and decelerated electrons encounter the catcher grids (on the output cavity), a strong and similar oscillation is created within that cavity area. As a result of delivering this energy to the output cavity, electron force on the plate is greatly reduced. The spend electrons are then removed by the collector plate (which is usually fitted with heat-dissipating fins). In order to inject and extract rf energy from the resonant cavities, small loops of wire are placed in each cavity.

A klystron may be operated as a microwave oscillator merely by connecting input and output cavities via a short length of coaxial cable. Coarse frequency tuning is accomplished by mechanically varying buncher to catcher grid distance, and fine frequency adjustments may be accomplished by varying the applied collector plate voltage. Since small frequency deviations are possible by these voltage variations, frequency modulation of the klystron's output signal can also be accomplished in this manner. This basic arrangement for producing an fm microwave signal reflects one of the klystron's major attractions when used as an oscillator.

The klystron may be used as an amplifier by applying an input signal to the buncher grid and extracting an increased, or amplified, version of that signal from the output cavity's catcher grid. In this particular case, no external connection is needed between input and output cavities.

Because the klystron's frequency is specifically affected by plate voltages, a well-regulated power supply must be used with these devices. Overlooking this fact will result in undesired frequency deviations and noise on the output signal. Most of the commonly available klystrons have a readily apparent frequency-tuning adjustment on the device's outer area. These particular units are relatively easy to get going at amateur frequencies.

MAGNETRON OPERATION

The magnetron tube is primarily employed for high rf power operations in the broad general range of 1 GHz to 5 GHz. These tubes are often used in various RADAR systems, therefore their surplus-market availability is reasonably good. If the amateur finds one of these tubes, he would be well advised to also obtain information and parameter details on its use. The tube is usually enclosed by one or two strong magnets (depending on particular magneron design). If the magnets have been subjected to extreme heat or sharp physical blows, their fields may be reduced to the point of rendering the tube useless.

The magnetron is essentially a diode device which operates on the principles of electron transit time, and the effect of a strong magnetic field on those electrons. This concept is shown in Fig. 2-6. Electrons in a conventional diode travel in a straight line from the cathode to the plate. When deflected by a strong magnetic field, however, the electron's path will bend to the point where it becomes circular. Resonant cavities are placed at the major points of these orbits, and the electron flow causes oscillations to be established within the cavities. The resultant microwave energy is then coupled to the outside world via a single loop place within one (or more) of the cavities. The cavity magnetron requires a balance of plate voltage, magnetic-field flux, and resonant-cavity tuning. Oscillations occur when these parameters are adjusted to a particular critical value.

The magnetron frequency like the klystron, may be adjusted

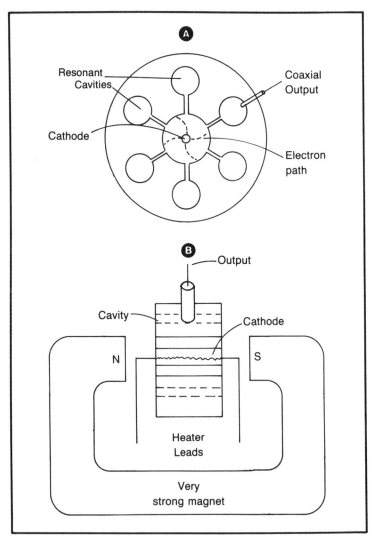

Fig. 2-6. Basic design of a cavity magnetron. Electrons are forced to spiral in their path by the strong magnetic field. As they pass the openings in the cavities, they excite the cavities, creating rf oscillations in the uhf or microwave spectrum. The view at (A) is a side view (with magnet omitted for clarity); (B) is an end view. Note that the magnetic field is parallel to the cathode.

by varying the resonant cavity tuning (area), and by varying plate voltage. Since electron activity is being coupled to a cavity, it's also possible to operate the cavities at frequencies harmonically related to the magnetron's fundamental range. Efficiency will be reduced in that case, however.

GUNN DIODE THEORY

The technique of generating low-amplitude microwave frequencies with solid state devices was discovered during the early 1960s by, appropriately enough, Mr. J. B. Gunn. working with a specially doped and diffusion-grown chip of Gallium Arsenide, Gunn found that when this device was subjected to a relatively low voltage, it produced a reasonably stable microwave signal in the range of 6 to 24 GHz. Additional research and development of the Gunn diode has improved its operation, and the device is now used in one of amateur radio's outstanding microwave units—the Gunnplexer. The Gunn diode proper is an extremely small device; it consists of two semiconductor layers having an overall thickness between 5 and 15 micrometers. Functions in a Gunn diode operate on the electron-transfer theory. Conducting at the speed of light, current pasing through the diode causes oscillation at a specifically established microwave frequency range. In addition to device layer thickness, physical mounts, and voltages impressed across the diode determine the operating frequency. In this respect, some Gunn diodes utilize a highly tapered body to permit smooth tuning over its frequency range. Product yield among Gunn diodes vary widely, requiring hand selection at manufacture for optimum results. Although the diode is a two-layer device, two additional layers (one forming a heat sink, open improving semiconductor material) are utilized to ensure acceptable performance an reliable life. Yet, with all the previously described elements, the Gunn diode is an extremely small device, typically 1/4 inch in finished form. A large number of these devices might be mistaken for mere fragments of metal.

As one may logically surmise, the extremely small Gunn-effect device is quite sensitive to excessive voltages. This restricting factor limits their microwave-energy output. Being a member of the Gallium-Arsenide family, the Gunn diode is a low-noise device. This aspect is also put to use in the form of 10-GHz local-oscillator injection. The Gunnplexer 10-GHz front end employs a Gunn diode placed within a 10-GHz cavity. A concentric or coaxial-type rf choke is used to connect power-supply voltages to the diode. While 10-GHz energy is directed from the cavity and radiated to the distant receiver, a small portion of that signal is also used as the local-oscillator injection. A set of Schottky diodes are mounted in the antenna Horn, and a ferrite-rod circulator is used to set the local-oscillator mixing level. The circulator couples approximately 10 percent of the Gunn diode's output to the mixer the remaining circuit. This, along with the incoming signal, produce a resultant i-f signal. Due to the required close tolerances and high quality of Gunn diodes, these devices are relatively expensive. Surplus-market purchasing, if possible, are strictly that, and it's quite doubtful if such diodes would be capable of providing acceptable results. Microwave Associates, Inc., secures their own top quality Gunn diodes, the relatively modest cost associated with complete and operational Gunnplexers is a very logical investment.

Chapter 3

Popular Microwave Bands

The three most popular amateur bands in the microwave spectrum are 23 cm (1,240 to 1300 MHz), 13 cm (2,300 to 2450 MHz), and 3 cm (10,000 to 10,500 MHz). The 23-cm band is presently quite active in most metropolitan areas of the world: ATV activity using the upper end, FM communications in the middle area and OSCAR amateur satellites occupying the lower end of this frequency allocation. A substantial amount of EME communications are also conducted in this range. While a miniscule amount of commercially manufactured equipment has been available for 23 cm, that situation is changing. Several noted manufacturers have geared up for this band, and the results of their endeavors should appear on the market around the time of this book's publication.

Even before amateur activity encompassed the full 23-cm spectrum, activity on 13 cm (2300 to 2450 MHz) began rising. Due to frequency stability and calibration requirements, the first operations were FM in nature. Today, however, stable circuits for 13 cm are being utilized for successful Amateur Television communications, amateur computer networking, etc. One of the more appealing, yet little known, aspects of this amateur band is its ability to provide comparatively inexpensive communications. This situation is due in part to the introduction of MDS equipment capable of operating in the nearby range of 2100 to 2200 MHz.

The methods of signal reception and processing begin to change form around 2 GHz, and techniques popularly known as downconversion involves receiving the signal and amplifying it as much as possible (and financially feasible) while holding inherent noise levels to a minimum acceptable level. Because 2-GHz active-device gain may be masked by its noise, the problem becomes a paramount consideration. A 10 dB gain with 7 dB noise, for example, has no advantage over a 4 dB gain with 3 dB noise. The noise situation simply must be overcome to acquire a desired 2-GHz signal. Following this critical 2-GHz rf amplification, the signal is heterodyned down to a lower frequency where it can be handled and processed by amplifiers with better signal-to-noise ratios.

The usual 13-cm downconverter is often placed directly at the antenna (which is often mounted in a resonant cavity). The downconverted signal is then passed via conventional coaxial cable to the i-f/processing setup. A one-pound coffee can has been found to serve well as a resonant cavity for 13 cm, and several companies are presently manufacturing downconverters that can be mounted on the end of these cans. The units are inexpensive, and they perform very well. An amateur who wants to operate fast-scan TV on 13 cm may thus employ a 2300-MHz transmitter and downconverter with his existing television receiver and become operational for a relatively low expenditure. A general outline of this arrangement is shown in Fig. 3-1.

Several manufacturers have recently begun producing transmitters for 2300 MHz, and their performance has proven very good. As little as one watt of power is sufficient for most line-of-sight paths, and the cost of such low-powered units is usually less than a bare bones 2- meter FM transceiver.

Although presently unconfirmed, the United States space shuttles are reported operating between 2200 and 2300 MHz during flights. The SWL challenge of receiving these transmissions is yet another inspiration for operating 13 cm. See Fig. 3-2.

If you're wondering whether amateur activity to a significant degree exists on 10 GHz, the answer is a resounding "yes!" Thanks primarily to the introduction of Gunn-effect diodes and the Microwave Associates Gunnplexer (actually a 10-GHz transceiving converter), activity is flourishing in this range. The narrow beamwidth and line-of-sight propagation at 3 cm allows simultaneous operation of numerous systems without interference; indeed, each user may be completely unaware of "neighbors" until duly informed. Such 10-GHz communications have often been compared to "invisible wires" linking amateurs. Low power is a fact of life at 10 GHz: 5 milliwatts being considered usual, and 15

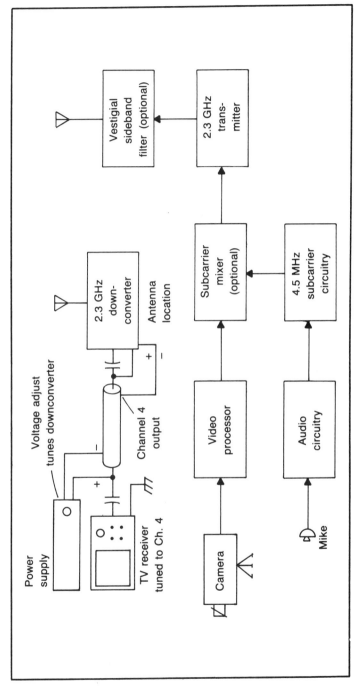

Fig. 3-1. A popular system for amateur fast-scan TV operation of 2.3 GHz. This setup is relatively inexpensive, but quite effective.

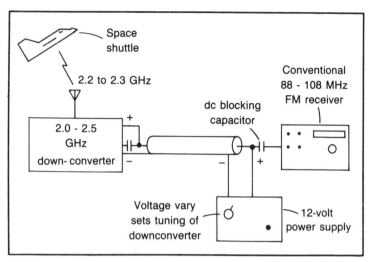

Fig. 3-2. Arrangement for using 2-GHz converter to receive transmissions from the space shuttle when it passes overhead.

milliwatts being considered "high power." Communication ranges are restricted by local terrain and obstacles, including heavy rainfall. Even with such limitations, amateurs have achieved communications via paths over 75 miles length on this challenging band.

The Gunnplexer is, in itself, an interesting and quite clever unit of very reasonable cost. The unit consists of Gunn diodes and Schottky mixer diodes mounted in a resonant cavity which is interfaced to a 17-db gain horn antenna. A photograph of the Gunnplexer is shown in Fig. 3-3. The Gunnplexer's rear section consists of a Gunn oscillator which converts dc into 10-GHz rf energy. Mechanical tuning of the cavity provides frequency shifts of up to

Fig. 3-3. The Microwave Associates 10-GHz Gunnplexer features a 17-dB gain horn that is mated to a cavity assembly that houses an oscillator and signal mixer.

100 MHz from the unit's nominal frequency. A varactor diode mounted close to the Gunn diode may also be used for frequency shifts up to 60 MHz, and for frequency modulating the transmitted 10-GHz signal. A Schottky diode is mounted near the horn/cavity junction area; it provides mixing action for reception of 10-GHz signals. During operation, the Gunn diode acts simultaneously as a transmitter and local oscillator for the receiving downconverter. A very small portion of the transmitted 10-GHz signal is coupled into the mixer diode, and a ferrite circulator is employed to isolate transmitter and receiver functions. Since a pair of communicating Gunnplexers are necessarily transmitting and receiving simultaneously, their frequencies are offset by the amount of the desired i-f. An example of this arrangement is illustrated in Fig. 3-4. The frequencies are offset by 146 MHz, and conventional 2-meter FM transceivers are employed for i-f stages. A small amount of 10-GHz energy from each Gunnplexer mixes with the incoming 10-GHz energy from the other Gunnplexer, producing an output of 146 MHz. It should be noted, also, that other i-f ranges could be used as well. Additional Gunnplexer information is presented later in this book.

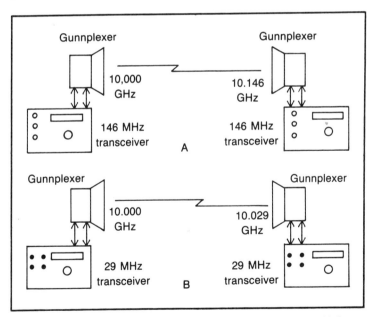

Fig. 3-4. Two means of using amateur FM transceivers in conjunction with Gunnplexers for 10-GHz communications. The transceivers serve as the i-f system when receiving and as the modulated injection source when transmitting.

Fig. 3-5. A 2-GHz converter constructed on double-clad printed-circuit board. The rear metal surface acts as a ground plane for stability.

CIRCUIT AND ANTENNAS FOR THE 13-cm BAND

As mentioned previously in this book, circuit design and layout at 2 GHz is quite different from that employed at lower radio frequencies. State-of-the-art designs center around the use of high quality G-10 double-clad printed-circuit board and low-loss/low-noise components. Since point to point wiring is virtually non-existent, stray capacitance is thus negligible. As a means of further clarifying construction/circuitry techniques for 13 cm, a typical downconverter rf unit is shown in Figs. 3-5, 3-6, and 3-7. The complete downconverter is constructed on one side of a double-clad pc board. The rear section is unetched, and serves as a ground plane to provide stability. The board is cut in an octagonal shape to fit the rear area of any one-pound coffee can. This type of feed provides approximately 11 dB gain over a basic antenna.

The downconverter's antenna connects via a short piece of miniature coaxial cable to the board's center top section; the shield connects to the rear ground plane and the center conductor protrudes through the board. A quarter-wave length match system is employed at the antenna input; each end of that strip being connected through the board to the rear-area ground. Since a quarter wavelength line exhibits impedance-inverting properties, an open circuit is thus reflected to the antenna connection point proper. An on-board etched capacitor couples signals to the rf amplifier which is mounted over a hole on the board to reduce lead length. (Placing the transistor on top of the board would require excessively longer leads.) Near the board's middle, mixer diodes are also mounted over holes to minimize excess inductance and capacitance. Note that extremely quick and accurate solder techniques are required, otherwise the glass diodes would be destroyed by heat. Barely visible near the diodes right strip line is a small leadless chip capacitor. The metallic strip along the board's left side connects all grounds on circuit side, plus connecting ground to the pc board's backplane. The left transistor (Q3, oscillator) connects to its associated stripline. The output from this stripline is coupled directly to its above area strip, which is directed to the mixer diode's left stripline. The signal difference (2154 MHz minus 2100 MHz)

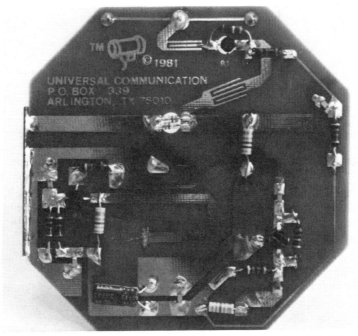

Fig. 3-6. A full view of the 2-GHz downconverter shown in Fig. 3-5. The rf amplifier is at the top, mixer diodes near the middle, local oscillator at the left bottom, and the i-f amplifier is at the right bottom.

35

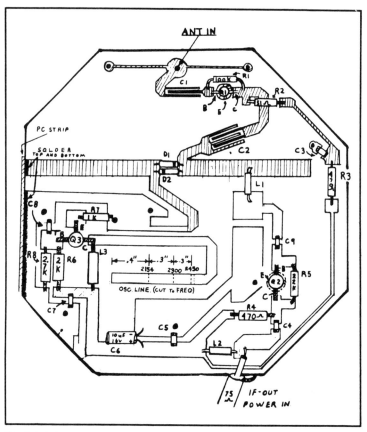

Fig. 3-7. A detailed drawing of a 2.3-GHz converter as shown in the photographs (courtesy Universal Communications).

is acquired from the right stripline's middle connection, passed through an encapsulated coil, a chip capacitor, and to the base of an i-f stage amplifier (Q2). The output signal from this device is then directed to the converter's bottom strip, where it feeds the indoor receiver via coaxial cable.

Tuning of the local oscillator stage is accomplished by varying stripline length. Lower frequencies require longer strips, and higher frequencies require shorter strips. The approximate tuning range of this strip ranges from 2000 to 2500 MHz.

The complete downconverter unit may be considered a front end that is used in conjunction with an external receiver. The downconverter's circuitry thus consists of its rf amplifier (top section), local oscillator (left section), twin-diode mixer (middle area) and a stage of i-f amplification (right bottom area). The complete unit is void of interconnecting wires; each component is placed at its proper location and soldered to its associated printed circuit strip. While construction of a 2-GHz downconverter may be accomplished without the use of rf amplification, its sensitivity would be quite low. Although the rf amplifier generates noise, the resultant acquired gain overrides that noise by a creditable amount. However, it is possible to bypass the rf amplifier and connect the antenna directly to the mixer. Its insertion point would be situated one-half wavelength from the diode locations. This point would reflect a direct connection to the mixer diodes. If great distances are not a concern, direct mixer-to-antenna connection is feasible.

Finally, sharp-eyed readers may ponder the existence of emitter leads for the i-f amplifier transistor (Q2). This transistor is also mounted over a hole, with emitter leads bent straight down, folded and soldered to the rear ground plane. The additional lands of solder around the board are grounding wires run through the board and soldered for additional low-inductance grounds.

The direct communications range at 2 GHz is primarily dependent on terrain, because, as previously mentioned, light of sight is necessary. High amounts of rf amplification seldom increase these distances substantially; however, they do provide more noise-free communications. Additional information on 2-GHz systems will be presented in a subsequent chapter.

DESIGNS FOR 3-cm EQUIPMENT

As one might logically surmise, circuit designs and construction techniques for 10 GHz are somewhat different from those employed at 2 GHz. Tuned-line tank circuits give way to resonant cavities, pc boards are eliminated, and components are mounted directly within cavities. Rather than delving into lengthy discussions of surplus microwave equipment, magnetrons, klystrons, etc. which may be modified for use in this amateur range, our discussion will be confined to present amateur state-of-the-art devices; namely the Microwave Associates Gunnplexer. This unit is so chosen because of its availability, simplicity, and relatively low cost. These units are dubbed "transceiver front ends" because they are used in conjunction with an hf/vhf transceiver that provides an i-f signal on both transmit and receive. Voltage is applied to the unit's internal Gunn diode through a resonant decoupling stub. Likewise, the i-f output signal is extracted by a tuned line/stub. These measures provide isolation of the 10-GHz signal from outside effects. Local oscillator and mixer actions happen inside the Gunnplexer cavity.

The Gunnplexer may be considered a totally independent 10 GHz signal source/10 GHz receiving converter. Two Gunnplexers may be used for communications by offsetting their transmitting frequencies by the amount of the desired i-f. As a result of both units transmitted signals "beating" in the mixer diodes, the resultant difference signal (i-f) is produced. It should be noted that i-f bandwidth of these 10 GHz Gunnplexers can be extremely broad; depending on applied signal bandwidths, i-f designs, etc. A portion of this signal loss may be compensated by high gain antennas. A basic outline of i-f bandwidth versus approximate range in miles is presented in Fig. 3-8.

In order to work over distances above 50 miles, bandwidths between 20 and 100 kHz are desirable. The prime consideration for these narrow bandwidths involves stable oscillator operation and consequent use of phase-locked-loop afc (automatic frequency control) systems. One example of such a system is shown in Fig. 3-9. The ability to hold Gunnplexer oscillator drift to less than 350 kHz per degree centrigrade when utilizing a 100- to 200-kHz band-

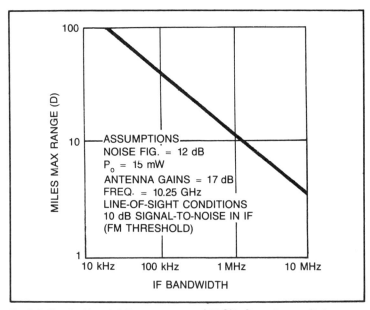

Fig. 3-8. Graph of bandwidth versus range of 10-GHz Gunnplexer units (courtesy Microwave Associates).

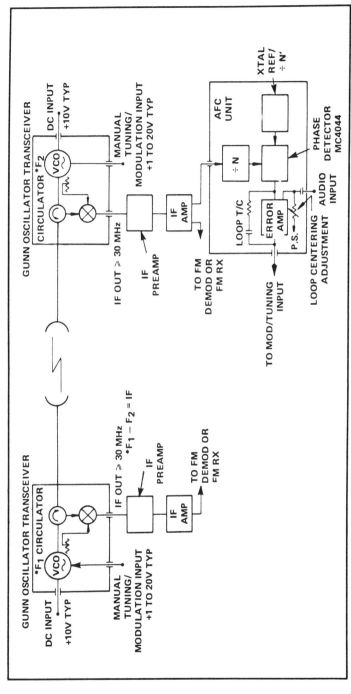

Fig. 3-9. A method of applying digital automatic frequency control (courtesy Microwave Associates).

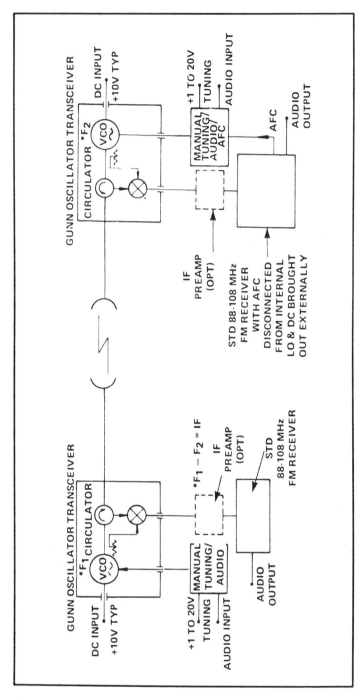

Fig. 3-10. An analog afc technique for phase-locking Gunnplexers as described in the text (courtesy Microwave Associates).

width requires a bit of ingenuity. Daily temperature changes of, for example, 25 degrees centigrade, can cause frequency shifts of nearly 9 MHz. Fortunately, however, Gunnplexers usually "settle into their environment" and reflect gradual frequency shifts during daily periods (low shifts per hour). However, with such gradual frequency drifts, the Gunnplexer's electrical tuning range of 60 MHz is quite adequate. These corrections may be accomplished by applying a voltage to the unit's varactor diode. In the setup in Fig. 3-9, one unit's vco is allowed to drift while the other unit's vco is set to track the proper i-f provided to the receiver. The i-f and a crystal-controlled oscillator are divided by N and the two outputs are frequency and phase compared. The resultant dc output is amplified and applied to the vco's varactor diode. The R/C shunting network shown is to prevent modulating signals from affecting the dc amplifier.

Another form of afc correction arrangement is shown in Fig. 3-10. A standard FM receiver (88 to 108 MHz) of good quality is used as an i-f amplifier. The FM receiver must be modified to disconnect its internal afc control from the internal local oscillator, and apply it instead to the Gunnplexer's varactor diode. The second transceiver also uses an 88 to 108 MHz FM receiver as an i-f system. Since the second vco is not afc corrected, it is merely tuned to the same frequency as the first unit and left unmodified. This setup allows the operator to correct manually either with the manual tuning control of the varactor-diode power supply, or by the frequency control of the FM receiver. The operator merely needs to ensure the vco with afc is set on the correct side of the other unit's vco so that frequency corrections converge rather than diverge. A final amount of fine tuning according to environmental conditions should bring the system into reliable operation over conventional line-of-sight distances. It should be kept in mind that this setup is not a superb DX arrangement, but rather an easily obtained method for 10-GHz operations. Details of a proven long-distance 10-GHz system are presented in Chapter 6.

Chapter 4

Communications Equipment for 1.2 GHz

This low microwave band offers a number of challenges to today's amateurs. Associated activities include OSCAR satellites, QRP, and amateur fast scan TV. We are quite honored to have an established uhf, microwave and ATV specialist, Tom O'Hara, W6ORG, presenting information from his area in this chapter. The following information is thus presented "intact," courtesy of W6ORG.

The amateur band of 1215 to 1300 MHz is actually the lowest authorized frequency range where there has been little commercially made equipment available and which gave the home constructor an opportunity to experiment with microwave techniques. With technological progress and parts-cost decreases, manufactured modules are appearing on the market that will allow more people to try this band by taking the systems approach.

This band was once occupied by just a few hams interested in CW and SSB DX centered around 1296 MHz. Today, there are quite a few ATV repeater outputs, ATV simplex and duplex stations, FM repeaters and relay links, plus several planned OSCAR satellites. To prevent mutual interference, and to let newcomers know where to look for those interested in a particular mode, a band plan was agreed upon.

23 cm BAND PLAN

1215-1239 MHz   Experimental, modulated oscillator, wideband data.

1240-1246 MHz	ATV simplex or duplex with 70 cm, 1241 video carrier.
1246-1252 MHz	FM relay & links.
1248-1258 MHz	Primary ATV repeater output, 1253-MHz video carrier.
1258-1264 MHz	FM relay & links.
1260-1270 MHz	Satellite uplink. ATV repeater output 1265-MHz.
1270-1276 MHz	FM relay & links.
1272-1282 MHz	ATV secondary repeater output, 1277-MHz video carrier.
1282-1288 MHz	FM relay & links.
1284-1294 MHz	ATV links, 1289-MHz video carrier.
1294-1295 MHz	FM relay & links.
1295-1297 MHz	CW, SSB & weak signal operation, 1296.0 calling frequency.
1297-1300 MHz	FM relay & links.

As of this writing, the 1979 WARC Treaty has not been fully ratified or implemented. There are two parts of this treaty which may affect the band plan slightly. First, the portion form 1215 to 1240 may be lost to radio navigation on an exclusive basis rather than a shared basis. Second, when a satellite goes into operation, existing repeaters or high-power systems in the 1260-1270 MHz segment may be moved. In this case, ATV simplex on 1241 MHz would swap with any ATV repeater outputs on 1265 MHz. If the bottom end of the band is also lost, then any ATV using 1241 MHz must use vestigial sideband filtering to stay within the band edge.

ATV and weak-signal-mode stations usually employ a 2-meter calling and coordinating frequency when operating the microwave bands. Some of the frequencies most used for that purpose are 146.43 or 144.45 MHz FM simplex. Because 2 meters usually gets out farther than the higher bands, one can monitor 2 meters with an omnidirectional antenna while waiting for an opening, rather than possibly missing it by having a highly directional 23-cm antenna pointing the wrong direction.

AVAILABLE EQUIPMENT

There doesn't seem to be any ready-to-go transceivers available at ham stores as there are for the 70-cm band, but the closest thing to it is the Spectrum International MMT1296-144 Transverter.

When mated with a 2-meter transceiver, this unit allows operating on most modes on 1296 MHz. It simply mixes and up-converts the two meter AM, CW, FM, or SSB energy to 1296 MHz, and provides about 3 watts output. In receive, it contains a 1296 MHZ to 2-meter downconverter. This is probably the easiest and quickest way to get on this band.

Spectrum International also has crystal-controlled downconverter modules to convert the 23-cm band down to 2 meters for voice and CW modes, or TV channels for ATV. For CW, AM, FM, or ATV, they have a model MMV1296 varactor tripler that accepts up to 25 watts input from a 70-cm transmitter, and multiplies it by three to the 23-cm band with 50 to 60 percent efficiency. AM modes work well if drive is kept to about 1/2 the maximum power capability of the varactor diode, and the circuit is tuned for best input/output linearity. Since the amplitude modulation envelope is the same as the input, the sidebands generated are also the same as the input, and are not tripled as is the case with FM deviation. SSB, however, does not work very well because even a small amount of intermodulation distortion from non-linearity will tend to reinsert the opposite sideband and carrier.

ATV has become the primary user of the rest of the band with five channels available. The channels, however, primarily are used as repeater outputs. Tuneable downconverters are available from P.C. Electronics which convert the whole band down to TV channel 7 or 8. The converter is mounted at the antenna to eliminate

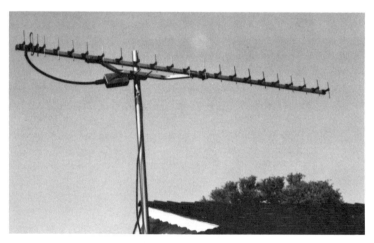

Fig. 4-1. An F9FT Tonna 23-element Yagi antenna with a PC Electronics TVC-12A ATV converter mounted on it.

Fig. 4-2. A 1253- or 1277-MHz ATV transmitter system designed by W6ORG. The transmitter is in the weatherproof housing at the right, designed to be mounted at the antenna. The control box is at the left.

the high feedline loss. See Fig. 4-1. This results in more sensitivity than if the converter is placed in the shack and a long lossy coax cable is run to the antenna. A variable voltage of 11 to 18 volts is fed up the 75-ohm RG-6 coax, and this voltage varies the downconverter's varicap-tuned oscillator. By transmitting to the repeater on 70 cm, and receiving on 23 cm, ATV stations can see their own video pictures coming back to them with no special filtering required. A minimum of five foot vertical antenna separation is usually sufficient to prevent receiver desensing during transmissions. This is a great benefit for lighting and camera adjustments, or running computer games and VCR tapes on the air. Some of the 70-cm inband repeaters have a secondary output on 1253 MHz which is used to transmit weather RADAR video, links to and from other repeaters, etc. Full duplex video and audio can also be run between two stations at the same time by one station on 426.25 MHz and the other on 1241 MHz.

The basic 10-watt ATV module package from P.C. Electronics is used in the 1253-MHz transmitter with the addition of a MMV1296 varactor tripler. The transmitter shown in Figs. 4-2 and 4-3 was built for use at the Jet Propulsion Lab in Pasadena, California, to retransmit the direct Saturn pictures from Voyager 1 and 2. It is built in a die-cast aluminum box for mounting at the antenna. The control box supplies 18 Vac power, video, and microphone audio through cables of the transmitter module. The 13 volt, 3

Fig. 4-3. Inside view of the W6ORG ATV transmitter. It uses a TXA5 exciter and a PA5 10-watt power module, driving a MMV1296 varactor tripler. The system also includes a FMA5 sound-subcarrier generator and a 14 Vdc power supply.

ampere regulator must be placed at the transmitter for sufficient regulation with ATV modulation. The module was also used as a portable 434 to 1277-MHz repeater for float coordination at the Pasadena rose parade. See Figs. 4-4 and 4-5.

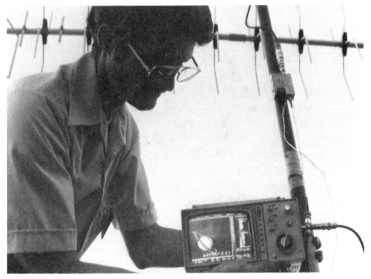

Fig. 4-4. Picture of the planet Saturn as repeated through the W6ORG/W6VIO 1277-MHz repeater and received by WB6BAP at the Griffiths Observatory.

Fig. 4-5. A portable repeater system using 434-MHz input and 1277-MHz ATV transmitter. This system was used for coverage of the Pasadena Rose Parade.

In a similar manner, an FM link or repeater transmitter can be made using a 70-cm transmitter and a MMV1296 varactor tripler. The deviation will, however, triple. The resultant 15-kHz output deviation might be acceptable for your system, or you can simply turn the 70-cm deviation control down until the usual 5-kHz output deviation is reached. A 2-meter FM receiver strip with a MMK-1296-144 crystal-controlled converter ahead of it will work fine for receiving.

P.C. Electronics makes small, one-milliwatt, modulated oscillators which work well for testing antennas and downconverters. It's an inexpensive signal source for the band, compared to a full signal generator. The TVG-12 can be modulated with audio or video. In fact, W6LLN built the TVG-12 into a small chassis and placed it above his color camera to use a "creepie peepie." It does a good job transmitting color video for about 1/4-mile around swap meets, parade coordination, etc., to a good gain antenna and receiver. A quarter-wave whip and ground plane serve as the transmitting antenna. Everything is powered from one 12-volt Gelcell battery. See Fig. 4-6.

23-cm DX

As the frequency increases, so do the path losses. If all else is equal (same antenna gain, distance, transmitter power, etc.) there is approximately 8 dB more free-space path loss on 23 cm than on

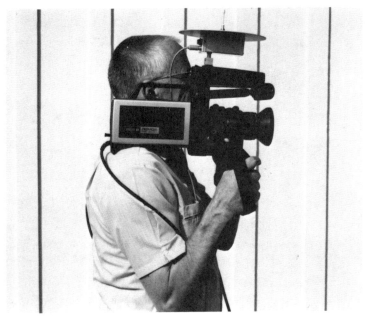

Fig. 4-6. Bob, W6LLN, with his 1241-MHz "Creepie-Peepie" color ATV system. Note the small ground-plane antenna above the camera.

70 cm. Since antenna physical size is 1/3 less, more gain can be put into the same antenna volume to make up for it. The F9FT Tonna 23-element Yagi antenna, for example, has a measured gain of 16.3 dB with only a 5-foot, 10-inch boom length. A two-foot dish with a dipole and splash plate reflector has about the same gain (16.3 dB). Free-space or line-of-sight calculations do not tell the whole story either. Temperature-inversion ducting, typical during the summer and fall months, send UHF and microwave signals far over the horizon.

A good example of duct DX is the KH6HME 1296 beacon at the 8200-foot level of the Mauna Loa volcano on the big island of Hawaii. It has been in operation since June of 1981. It was heard in Garden Grove, California, by W6KGS (DX of 2500.8 miles) and Harbor City, California, by K6ZMW on the evenings of July 30 and August 10, 1982. At times, the signal was as high as 36 dB above the noise. Unfortunately KH6HME was visiting California during the opening, otherwise there would have been good two-way contacts. The two-way DX record is thus still held by the Australians. See Fig. 4-7.

Fig. 4-7. The antenna system for the 1296-MHz beacon at KH6HME in Hawaii consists of four loop Yagi antennas.

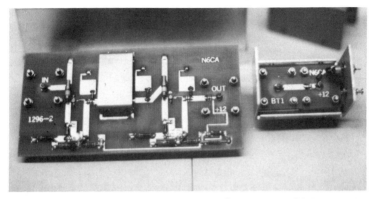

Fig. 4-8. A 1296-MHz preamplifier designed by Chip Angle, N6CA. It uses stripline construction and dual NE64535 devices.

The KH6HME equipment was designed and built by Chip Angle, N6CA. It can operate in two modes: beacon and radio. A 28-MHz i-f signal is up converted to 1296 MHz, and amplified to one watt by solid-state amplifiers. The final power amplifier is a single, air-cooled, 7289 triode cavity amplifier giving 25 watts output. The receiver has a 0.5 dB-noise-figure GaAsFet preamp ahead of two NE64535 stages. Received signals are then downconverted to a 28-MHz i-f strip. Two coax relays are used for antenna switching: one for protection of the preamp, and the other for actual switching between the receiver and the transmitter. A 20-foot piece of 7/8-inch Heliax hard line connects the equipment to the four vertically stacked, horizontally polarized, 25-element loop yagis. The array is pointed at Point Conception, California, but at that distance, the 3-dB beam-width points are at San Fancisco and San Diego. See Fig. 4-8.

The 23 cm band is an excellent band for the beginner, the builder, ATVer, or microwave DXer. We encourage you to investigate this unique "low microwave range."

Chapter 5

Communications Equipment for 2.3 GHz

One of the outstanding aspects of the amateur 2.3-GHz range is its rather extensive bandwidth. This permits communications of various natures ranging from basic voice modes to transmission and reception of broadband, high-resolution, video signals. Although these bandwidth allocations are a basic lore of the 2-GHz band, they are far from the only reason for amateur experimentation and pioneering there. Communications in the 2.3-GHz range are relatively free of interference and, due to their line-of-sight propagation, quite predictable in their actions. A certain amount of privacy is also inherent. Obviously, the amateur oriented toward microwave pioneering would reap maximum rewards by beginning with a basic narrowband voice system, and later expanding the system to include more exotic modes of operation.

The 13-cm band is a relatively cost-effective range for microwave activities. Amateur 2.3-GHz transmitters, complete with etched tuned-circuit strips, can be assembled on printed circuit boards at low cost, and other common items can be pressed into service as antennas. One-pound coffee cans, three-inch metal washers, and children's snow scooters, for example, can be effectively used in antenna designs for 2.3 GHz. Popular converters make reception of 2.3-GHz signals quite simple.

Another outstanding aspect of the 2.3-GHz range involves expanding capabilities as the enthusiast acquires microwave operating finesse. Starting with a simple audio system lacking any form of automatic-frequency control, an operator can add an afc circuit to the basic system, to achieve improved communications. Next, a phase-locked-loop form of frequency control might be added. Subsequent expansions might then include television links, computer interconnects, etc. Such systems could provide capability which, through the use of narrowband filters and precise frequency control, would raise the brow of many commercial operations. This chapter will present some general outlines in that direction. Additionally, information about available 2.3-GHz equipment will be presented.

SETTING UP A 2.3-GHz AMATEUR SYSTEM

A straightforward communications setup for 2.3 GHz can be constructed around the usual items of a transmitter, receiver and two antennas. When dealing with these higher-than-usual frequencies, however, a few minor alterations are required. Separate antennas for the transmitter and receiver are desirable, because transmit/receive switching at these short wavelengths is difficult and lossy. Circulator-type arrangements may be used for signal steering, their initial design and assembly by the microwave newcomer is not advised.

The receiver is usually separated into two items: the antenna-mounted converter, and the indoor i-f/tuning unit. Likewise, the transmitter is usually mounted at the antenna, with power, tuning, and modulating signals coming from the indoor source. A general outline of this arrangement is illustrated in Fig. 5-1.

Most popular 2.3-GHz converters are tuned by adjustment of their on-board strip line. The i-f output range is then adjusted by varying the operating voltage to the converter. This range is usually adjustable from approximately 52 to 90 MHz. A commercial FM receiver connected to the converter's output is considered a good starting point for this microwave activity: signal levels will be acceptable and initial expense will be low. This arrangement, however, isn't without minor consequences. Most 88 to 108-MHz FM receivers have bandwidths of approximately 150 MHz. An amateur 2.3-GHz FM signal may be confined to less than 50 MHz, thus the extra bandpass will be filled with signal-masking noise. Alternative solutions involve using an excessive amount of modulation to fill the extra passband (undesirable), or employing a narrow-band receiver following the converter (preferred). Such receivers may be obtained from government surplus outlets, or amateur 50-MHz

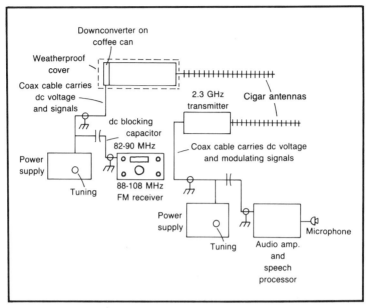

Fig. 5-1. An amateur transceiving setup for voice communications on 2.3 GHz. Although a conventional FM receiver is shown for reception, a more narrow-band i-f system may be used.

gear may be modified for reception in this range. There is nothing mysterious or unique in the 54- to 90-MHz frequency range of a converter's output. Any receiver connected to that source will reproduce signals—along with any noise appearing in the unoccupied passband.

The 2.3-GHz transmitter shown in Fig. 5-1 is typical of those available from sources such as Universal Communications, P.O. Box 339, Arlington, Texas 76010. These units are usually available in the basic form of 100 milliwatts, with add-on power modules to provide up to 10 watts output. As this book is being written, 2.3-GHz transmitter kits are relatively scarce. This situation may change during coming years, however, as amateurs acquire the necessary specialized construction and tuning techniques.

Assuming two or more amateurs establish an audio system as previously outlined, the next move will probably entail adding higher gain stages, larger antennas, etc., for increasing communications range and DX ability. Next, the desire for video or computer interlinking may be felt and the system will begin to grow. Fortunately, the necessary bandwidth requirements are ready and waiting for such expansions.

EXPANDING THE 2.3-GHz SYSTEM

The previously described bare bones audio setup for 2.3 GHz can be expanded to include amateur fast-scan TV activities in a relatively simple manner. Assuming one possesses the essential items for FSTV operations (a black and white or color camera and a conventional TV receiver or closed-circuit monitor), the setup can be rigged in a short time. The video modulating signal is directed to the transmitter's input, and a television receiver is connected to the downconverter's output. Due to the use of video frequencies approaching 4 MHz, the camera-to-transmitter cable length should be held to less than 70 feet. If this arrangement is not feasible, the video signal may be conveyed to the 2.3-GHz transmitter via an alternate frequency low-power link. The baseband video is then removed from that receiver's video detector/amplifier and directed to the 23-GHz transmitter. Because transmitter output is directly affected by modulating-signal bandwidth, the TV signal will have less power than its audio predecessor. Therefore a power amplifier is suggested. Because the converter output frequency range is relatively broad, a slight readjustment of operating voltage is all that's required to move the i-f range from approximately 90 MHz (audio reception with use of conventional FM radio) to approximately 82 to 88 MHz (frequency range for TV channel 6). The next improvement might include combining a 4.5-MHz sound subcarrier to this setup. One suitable audio-subcarrier generator is available from P. C. Electronics.

This unit is very easy to install and operate; the audio modulation is applied to one of the inputs, and video is applied to the other input. The resultant video with subcarrier output is then directed to the microwave transmitter. The received signal is tuned in the same as any conventional television signal, and the results appear professional and quite impressive.

Can the amateur TV microwave setup be used effectively for color TV operations? Indeed it can, and the outcome is truly spectacular. Contrary to first impressions, color TV activities are not extremely complex or expensive; the only required items are a color CCTV camera and a color television receiver (home entertainment-type cameras are presently marketed for under 600 dollars). Setup of the color system is, again, quite simple: the camera's output is directed to the transmitter and the converter's output is directed to the color receiver. See Fig. 5-2. Since bandwidth considerations associated with color TV operations are important (an almost full 6 MHz of spectrum is desirable), a minor decrease in communications range can be expected. Higher power levels will, again, overcome that annoyance.

Home-computer linking via 2.3 GHz is also relatively easy to achieve. The associated terminals are connected to their respective modems, and the resultant signals are connected to the microwave communications system. By using different transmitted frequencies, or a multichannel subcarrier, many computer terminals may be connected to a single microwave transmitter or receiver. The computer's baseband video signals may be relayed

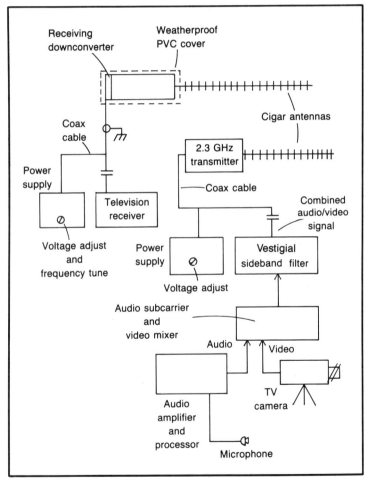

Fig. 5-2. An amateur setup for television operation on 2.3 GHz. The vestigial sideband filter and audio subcarrier are optional. Either black-and-white or color can be used.

via 2.3 GHz, rather than by conventional modem signals. Many home computers provide such raw video outputs. It is the signal that is applied to a vhf-TV modulated oscillator to view computer functions on a TV set. One may thus apply a computer's raw video to the microwave system and view the results either locally or at a distance via a conventional TV set connected to a 2.3-GHz downconverter.

QRP AT 2.3 GHz

Considering the typically low rf power levels of most 2.3-GHz systems, the term "QRP levels" might seem somewhat humorous. Such is not the situation, however, because those low-powered transmitters hold some unique capabilities for the experimenter. Through the use of a miniature 2.3-GHz transmitter, for example, a truly wireless amateur television camera is possible. The transmitter proper may be placed inside the TV camera, and an associated 1.1-inch, quarter-wave whip can be extended from the camera's top for video transmission. Although this setup exhibits extremely low range (500 to 1000 feet typical, at 1 milliwatt of power), its capabilities when used with a portable camera/VCR are unique. Low-power 2-GHz transmitters are also useful for in-house video relays such as TV games, computer displays, etc. A home computer may be set up with its associated video display in one area, while a very small and self-contained 2.3-GHz transmitter also relays video displays or readouts which can be tuned and received with a simple downconverter connected to any unmodified television receiver. Pursuing this a step further, amateurs located in almost adjacent areas might employ such QRP links for combined use of a computer or digital-scan converter. Assuming, for example, several SSTV enthusiasts secure a slow-scan-television scan converter, a pair of 2-GHz links can be used in lieu of hardwiring to the (digital) unit's inputs and outputs. As an end result of this arrangement, the remote digital scan converter would be operating on a round-robin format serving all operators. Numerous other capabilities are possible—the expansions and possibilities are limited only by one's imagination.

One of the most appealing and functional 2-GHz QRP transmitters presently available is the TVG-23 2300-MHz ATV test generator/transmitter being produced by Tom O'Hara, W6ORG, of P. C. Electronics in Arcadia, California. The schematic diagram of this unit is shown in Fig. 5-3, and a photograph of the circuit

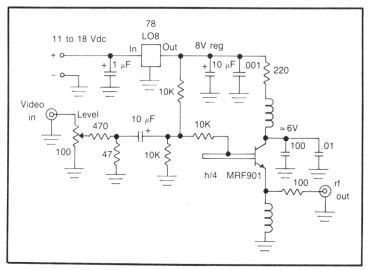

Fig. 5-3. Schematic diagram of the W6ORG low-power 2.3-GHz transmitter described in the text.

board is included in Fig. 5-4. Our thanks to Tom O'Hara, W6ORG, for sharing this information with us.

The TVG-23 ATV test generator is a simple video-modulated, free-running stripline oscillator for use in adjusting converters, or as a QRP 2375-MHz ATV transmitter. Power requirement is 11 to 18 Vdc at 15 mA. Any standard video source from TV cameras, VCRs, color bar-and-pattern generators, or even audio, as long as it is between 0.5 and 2 volts peak to peak into the 75-ohm load. The 100-ohm potentiometer in Fig. 5-3 adjusts the modulation level.

Frequency range is approximately 2100 to 2500 MHz. At microwave frequencies, it takes very little stray capacitance to greatly change the frequency. The base-bias resistor lead alone has enough capacitance to the ground plane to change the frequency

Fig. 5-4. Photograph of the low-power 2.3-GHz transmitter described in the text.

as much as 80 MHz by moving it 1/8-inch along the quarter-wave stripline. The frequency increases as the lead is moved toward the oscillator transistor. To raise the frequency higher, say to 2375 MHz, a cut (with an X-acto knife) can be made at the end of the stripline, 0.1 inch from the end.

If the cut has been made for the higher frequency but you want to be able to tune lower again and back, solder a small piece of Teflon-insulated No. 22 solid wire 0.3-inch long, to the ground plane next to the end of the base quarter-wave line. The closer the wire is to the line, the lower the frequency. The rf output loading will also have some affect on the operating frequency, so make the adjustments after the unit is in any enclosure, and with the antenna or 50-ohm load connected.

Power output is about one or two milliwatts which is not enough to radiate more than about 200 feet from the bench without an antenna. This is fine for rough tuning a converter front end. For finer adjustments, the TVG-23 should be put in a well-shielded box with the power and video input fed in with feedthrough capacitors. The output can be via a chassis connector that can go to a step attenuator to vary the signal level. The power supply can be a small 12-Vdc unit, or a battery.

If an antenna is used, be sure the frequency is within the 2300- to 2450-MHz ham band to avoid possible interference with MDS transmissions on 2154 MHz, and other services on 2200-2300 MHz. Tune the signal in on a converter and receiver of known accuracy, a spectrum analyzer, or wave meter to make sure of its operation.

When tuning in the video, adjust for best picture. As with any modulated oscillator, there will be some FM along with the AM. You will also note some fluctuation while moving around the room due to multipath bounce off metal objects. Have fun with QRP at 2.3 GHz!

ANTENNAS FOR 2.3 GHz

The short wavelength of 2.3-GHz signals permits utilization of many unique and high-gain types of antenna. The basic 2-GHz dipole is seldom employed in its individual form. High-gain arrays are easily constructed (and vital to long distance propagation) at this frequency. Likewise, resonant cavities (such as a one-pound coffee can) are often used for achieving gain and for mounting the antenna's driven element proper.

The simplest and most basic form of antenna setup for 2.3 GHz uses an empty one-pound coffee can. The physical dimensions of

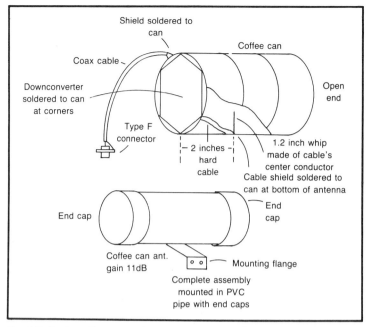

Fig. 5-5. Basic coffee-can antenna arrangement often used on 2.3 GHz.

these cans makes them resonant close to 2.3 GHz, and a downconverter can be affixed to the can's rear section. A 1/2- or 5/8-wavelength whip antenna is then placed inside the can and connected to the downconverter via a short length of miniature coaxial cable. See Fig. 5-5. Approximately 11 dB gain is provided by this configuration.

Cigar Antenna

The "cigar" antenna is a popular and widely used item on 2 GHz. A sketch is shown in Figs. 5-6, and 5-7. The front (director) section is approximately 35 inches long and consists of approximately 32 washers, 1-3/4 inches in diameter spaced 1-1/16 inches on a long section of rod or screw stock. Large nuts on each end hold the assembly together. A dual washer and nut arrangement is used for mounting the long section to a section of PVC pipe in which the downconverter and its associated coffee can are mounted. A drain hole is usually included in the PVC to prevent moisture condensation (which would rapidly rust the tin can). The PVC-to-coffee-can fit is rather tight. End caps are used with the PVC pipe. White is highly desirable, as darker colors can slightly attenuate

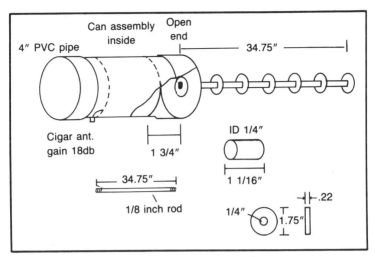

Fig. 5-6. The cigar antenna is an outstanding 2.3-GHz performer. The converter and its coffee-can mount are at the rear of the antenna, enclosed in PVC pipe for weatherproofing.

signals. Note that the longer washer section is not electrically connected to the resonant-cavity-mounted antenna; signal coupling is accomplished through electrostatic and electromagnetic fields. The cigar antenna, complete with PVC and coffee-can-mounted

Fig. 5-7. Author Dave Ingram, K4TWJ, making a final check of the cigar antenna and the converter before mounting it on tower.

downconverter provides a full setup for 2-GHz reception. Its approximate gain is 18 dB—a reasonably acceptable figure for this portion of the microwave spectrum.

Funnel Antenna

The funnel antenna consists of a section of fine wire mesh or extruded aluminum placed in front of a one-pound coffee can, as shown in Fig. 5-8. This arrangement provides approximately 16 to 18 dB gain, while acting as a non-polarized feedhorn for the coffee-can cavity. The funnel's wave-guide front area is 18 inches in diameter; the funnel's length is 23 inches, and the rear section is precisely the diameter of the one-pound coffee can. This relatively efficient antenna is easily constructed. Remember to provide electrical continuity between the funnel section and the coffee can, scraping to bare metal when required. Remember also to protect the downconverter and coffee can from weather exposure.

Parabolic Dish Antenna

The parabolic dish is unquestionably the top antenna for the 2-GHz microwave band. The larger this antenna, the greater its gain. The ever popular "Snow Scooters" or "Snow Coasters" have been found to be perfectly suited for use in the 2- to 2.5-GHz region. These items are available from large department, sporting-goods, or hardware stores. Their price usually ranges from eight to twelve dollars. Two types of Snow Scooters have been noted: A plastic model and an aluminum model. Obviously, only the aluminum one will work. The Snow Coaster is modified by removing its handles, and installing three pieces of aluminum angle bracket or small tubing flattened at the ends. These three pieces can be brought together to form a tripod, and bolted to a mount

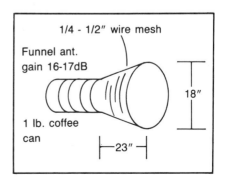

Fig. 5-8. The funnel antenna provides gain and directivity on 2.3 GHz in a small package.

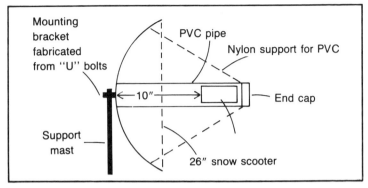

Fig. 5-9. The popular snow scooter serves as a ready-made dish for microwave work. It is easily mated with a PVC-enclosed converter and mounted on a mast. Gain is approximately 18 dB. Use thin-wall PVC pipe to save weight in the assembly.

which then holds the coffee-can assembly. Centering of the coffee can may be checked by placing the dish on a flat surface and holding a plumb bob to the coffee can's inner surface. Alternatively, a section of PVC tubing may be mounted to the Snow Scooter's center and used to enclose the coffee can and downconverter assembly. A sketch of this parabolic dish arrangement is shown in Fig. 5-9.

As previously mentioned, dishes of larger area may be used for 2 GHz. The dish need not be completely solid at this "low" microwave range. If screening or aluminum mesh is used, however, openings should not exceed 1/4 inch. Rather than delving into extensive mathematical calculations for determining focal lengths, simply move the coffee can forward and backward with respect to the dish until maximum output from a known, and hopefully remote, signal source is achieved.

Transmitters and receivers designed for operation on the amateur 2.3-GHz range are advertised each month in the amateur-radio magazines. Likewise, a variety of antennas and other unique devices for 2-GHz communications are advertised. The 2-GHz range is a very popular amateur frontier. Expansions and developments here can only grow during the coming years, proving the unlimited capabilities of what was considered a few years ago a mere line-of-sight band. Low-power communications on 2 GHz are a way of life, yet this energy can prove its worthiness—particularly for signals. The 2-GHz band is an ideal place for an amateur to begin microwave experimentation and operation. As knowledge is acquired and operating abilities improve, higher frequency bands can be explored with an almost guaranteed high level of success.

Chapter 6

Communications Equipment for 10 GHz

The 10-GHz band could easily become one of amateur radio's more popular allocations during the coming decade. Boasting a substantial spectrum allocation and capability for reliable line-of-sight communications, this range is destined to convey such various forms of advanced communications as multispectrum video, computer data-packet transfers, etc. As the amateur-radio population increases and technology for the microwave spectrum expands, this band will provide the dependable communications previously left to chance on lower frequency bands. Naturally, a full complement of amateur satellites will be necessary to fulfill this situation; however, plans for these satellites are under development at this time. Meanwhile, the amateur can begin getting his feet wet in this challenging area with presently available equipment such as Gunnplexers by Microwave Associates and associated support modules by Advanced Receiver Research.

As outlined in previous chapters, the general concepts used for 10-GHz operation involves using transmitting/receiving converters that permit using efficient i-f stages for signal reception and processing. The frequency range of this i-f is determined by the associated support equipment, and the 10-GHz front end is tuned via both mechanical and electrical adjustments. Any i-f range between audio frequencies and approximately 150 MHz may be used. However, ranges between 28 and 150 MHz are preferred because of their ability to include frequency-lock circuits. Due to the possibility of intermodulation and birdies from aircraft instrument landing systems in the 108- to 112-MHz range, that range should be avoided. The 30-MHz range is highly desirable. The Gunnplexer and its support modules are thus usually pretuned to this range; that is, one unit transmits on 10.250 GHz while the other unit transmits on 10.280 GHz. This form of operation is acquiring widespread popularity. A series of microwave links may also be established by "staggering" the i-f of repeating units: one on 10.250 GHz, one on 10.280 GHz, one on 10.250 GHz, one on 10.280 GHz, etc.

Frequency stability at 10 GHz is a major consideration for long-distance communications. In addition to fluctuations caused by variations of ambient temperatures, changes in power-supply voltages can also cause shifts in frequencies; shifts which may ultimately cause the selected i-f range to be missed. The obvious solution to this problem involves using well-regulated power supplies and basic temperature-stabilizing arrangements. Quite often, thermal stabilizers take the form of simple heaters or light bulbs placed (along with their thermostats) in proximity to the Gunnplexer. Fortunately, Gunnplexers usually establish equilibrium with their environment and drift rather slowly with day and evening temperature changes.

A BEGINNER'S SETUP FOR 10 GHz

An individual or a pair of amateurs can purchase their respective Gunnplexer and begin limited 10-GHz operations within a few hours time. Granted, such setups are far from optimum and useful for only a few miles, but they provide a convenient means of checking microwave equipment and acquiring an introduction to this band. This first setup resembles police radar-gun techniques while affording some interesting opportunities for the isolated microwave enthusiast.

A quick setup for experimenting with a single Gunnplexer can be rigged by the circuit shown in Fig. 6-1. Since the Gunnplexer will function successfully when potentials between 8 and 11 volts are applied to its Gunn diode, a 9-volt battery is thus used as the power source. This also eliminates the need for voltage regulation and large filtering capacitance. A second battery supply of approximately 3 volts is connected to the Gunnplexer's varactor diode input for establishing a specific operating frequency and for applying audio frequency modulation to the unit. Although a set frequency

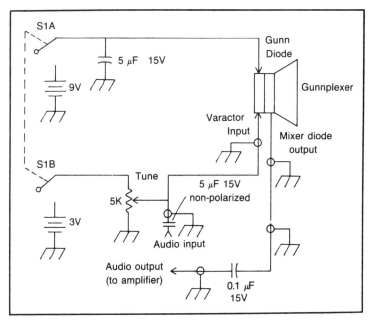

Fig. 6-1. A basic Gunnplexer setup as described in the text. Resistance and capacitance values are approximate, and not critical.

may not be necessary, this voltage is required for stable operation and for modulation purposes. An audio-signal generator or tape recorder (with taped tones) is then used as a signal source, and a conventional audio amplifier is used at the unit's i-f output port. The completed setup may be as "professional" or "haywired" as desired; just keep leads short as possible.

CAUTION: <u>Never peer into the output slit of a transmitting Gunnplexer—regardless of its output!</u> Eye tissues are extremely delicate. Likewise, bodily exposure to any level of microwave energy should be avoided. High levels of this energy (such as those encountered in radar systems) may destroy bone marrow, ultimately causing a form of leukemia.

The arrangement of Fig. 6-1 may be used in the following manner. First, connect the audio source output to the audio amplifier input and adjust levels as required to obtain proper volume. Next, disconnect the audio source from the amplifier, and reconnect each to the Gunnplexer circuit. Finally, double check all voltages and polarities (lest we accidentally damage the prized Gunnplexer at this critical period), and ensure the Gunnplexer is facing an unobstructed area for transmission. Apply power to the varactor diode and then the Gunn diode. If all's working correctly, the modulating tone will be heard weakly from the audio amplifier on the mixer output. Select a large metallic object (auto, metal building, etc.) at least 75 feet from the Gunnplexer and point the unit *precisely* in that direction. The transmitted beamwidth is quite narrow, thus slow "aiming" movements are suggested. As the 10-GHz signal reflects off the metallic object and returns to the Gunnplexer, an increase in tone amplitude will be noted. This effect is most apparent when low signal levels are used; the ear is more sensitive to such low levels. Alternatively, a vtvm or oscilloscope may be used to register signal-level variations. With a few minutes of experimentation, you'll become relatively familiar with Gunnplexer action, beamwidths, range variations, etc. Often, the Gunnplexer will be able to "see" better than its user: autos a few miles away, metal buildings obscured by small trees, etc., can be spotted. The Gunnplexer sees equally well in the darkness of night. Indeed, this simple setup can serve as a useful intruder watch in many installations.

The "radar-gun" concept can be tried next if desired. If the transmitting unit is pointed at a moving car, doppler shift will change the received frequency according to the auto's speed. Police radar guns employ this exact technique, with a differential amplifier, frequency counter (calibrated in miles-per-hour), and a "hold" circuit for measuring speeds. Try clocking airplanes if you like, but remember beamwidth is narrow and reflected signals and doppler shift occur only while the 10-GHz signal hits its target. Enterprising amateurs may also try reflecting 10-GHz signals off distant thunderclouds or lakes; the results are quite interesting. Measurements of distance in this case can be accomplished with the aid of a known time base and triggered sweep oscilloscope. The possibilities are endless! An amateur can easily become quite involved with the "singular form" of microwave activities, so we had best move on to communications setups before such bugs "bite" heavily!

A QUICK AND EASY 10-GHz COMMUNICATIONS SETUP

The "bare bones" setup of Fig. 6-1 requires only minor modifications for use as a two-way system. This arrangement is illustrated in Fig. 6-2. The previously used audio amplifier is switched to operate with a microphone for modulating the Gunnplexer, and a conventional 88- to 108-MHz FM receiver is used as a tunable i-f strip, detector, and audio amplifier. The varactor diode tuning-voltage supply is changed from 3 volts to 18 volts to permit electrical rather than mechanical tuning for this relatively high i-f. Whenever possible, mechanical adjustments to the Gunnplexer should be avoided (especially when accurate frequency- and power-measuring devices are not available). Bear in mind this is not a first-class setup, but rather a simple and effective means of getting operational on 10 GHz within a few hours. The effective working range, after both Gunnplexers are energized for a few minutes and allowed to establish equilibrium with their environment, will probably be a few miles.

Assembly of this quick and easy Gunnplexer system can be accomplished in either of two ways: simple or sophisticated. As this arrangement will probably be used as a stepping stone in your

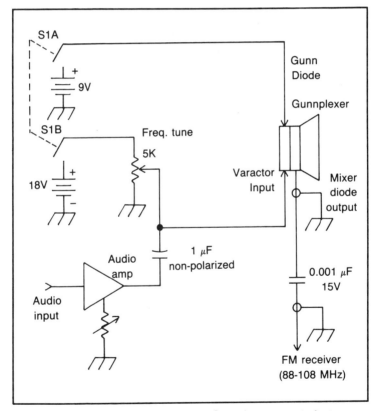

Fig. 6-2. A quick and easy setup to use a Gunnplexer as part of a two-way system. Systems of this nature can be put together quickly, and will provide communications over several miles. Any low-level audio amplifier may be used for the modulation stage (uA741, etc.).

microwave activities, the simple approach is highly recommended. Place the Gunnplexer on the front portion of a support cabinet and secure it with plastic or electrical tape. Next, place the FM receiver, batteries, etc., inside the cabinet and secure them with electrical tape. Finally, run a length of shielded cable to the microphone, allowing ample length for movements during use. Use small shielded cable to connect the Gunnplexer's mixer output to the FM receiver, etc., and all should perform satisfactorily when initially turned on.

Adjustment of the system is quite simple; one unit is held on a set frequency while the other unit is tuned for a signal. A 2-meter liaison link may be of great help during this initial tryout. Several variables are present when tuning this setup, consequently a logical pattern of adjustments is necessary to minimize confusion and prevent unnecessary fumbling (a certain amount of fumbling, however, is to be expected). As a starting point, set the varactor-diode voltage on both units at approximately 6 volts, and set both FM receivers on approximately 100 MHz. Next, apply 10 volts to both Gunn diodes and establish communications via the 2-meter link while both units achieve temperature stability. After a few minutes of settling down, tune the FM receiver connected to one Gunnplexer across its range and listen for a 10-GHz signal. This test first should be performed at a range between 1/4 and 1/2 mile to ensure "first-time success." If the 10-GHz signal transmitted by one unit cannot be received by the other unit's FM receiver, then (and only then) readjust the transmitting unit's Gunnplexer's varactor-diode voltage to 15 volts. Again, tune the other unit's receiver over the FM spectrum. If at this time the transmitted message still cannot be received, readjust the receiving Gunnplexer's varactor- diode voltage to approximately 15 volts, and retune its FM receiver.

This technique of holding one transmit frequency constant while adjusting the other Gunnplexer's receiver has been highly effective for achieving first-time results. If you have no absolute success, return to the initial setup described in the first part of this chapter to ensure proper operation of each Gunnplexer. Once communication has been established between Gunnplexers, however, distances can be expanded. Eventually you will get a feel for maximum operating range and the effect of line-of-sight versus non-line-of-sight paths on this band. If additional range is desired, the following setups are heartily suggested. Their track record is quite admirable!

A HIGH-QUALITY 10 GHz COMMUNICATIONS SETUP

This setup is constructed around the Advanced Receiver Research model RXMR30VD support system designed specifically for use with the Microwave Associates Gunnplexer. It features a high quality 30-MHz i-f receiver section, state-of-the-art discriminator and afc, plus transmit and receive audio sections. Essentially, this Advanced Receiver Research unit and a 10-GHz Gunnplexer (plus a few outboard switches, meters, etc.), provide a complete and reliable 10-GHz communication system. These modules are directly available to the radio amateur from Advanced Receiver Research, Box 1242, Burlington, Connecticut 06013. An amateur with limited available time and resources may thus become operational on 10 GHz in minimum time through the purchase of these items. Additionally, this setup reflects a logical and cost-effective method for assembling a quality 10-GHz station.

The Advanced Receiver Research model RXMR30VD system is supplied as an assembled, tested, and completely aligned unit built on a 3- by 6-inch circuit board. It includes the 30-MHz FM receiver, diode-switched i-f filters, dual-polarity afc system, diode-supply regulator, and modulators for either phone or CW operation. The unit is equally suited for fixed, portable, or mobile operation: a definite plus for serious 10-GHz operators. Power requirements, including current drawn by the Gunnplexer, are 13

Table 6-1. Notes for Fig. 6-3.

Note 1—The values of R1 and R2 will be determined by the type of meters used with the RXMR30VD. The values provided on the board are suitable for use with Modutec panel meters. A 0-300 uA meter is used for signal strength and a 0-1 mA for discriminator/manual tuning voltage. If other meters are used, some changes in R1 and R2 may be necessary. R1 is determined by injecting a 10 mV signal into the receiver and selecting a value for R1 that causes the meter to read full scale. R2 can be determined by placing S1 in the discriminator position and selecting a value for R2 that will cause the meter to read mid scale.

Note 2—S2 and S3 may be condensed into a single switch - dpdt with center off.

Note 3—If a squelch control is not desired pin 19 should be connected to ground.

Note 4—As supplied, the RXMR30VD is designed for a high-impedance microphone. For use with low-impedance microphones, simply connect a 1/4-watt resistor the value of which equals the characteristic impedance of the microphone across the microphone connector or from pin 4 to ground.

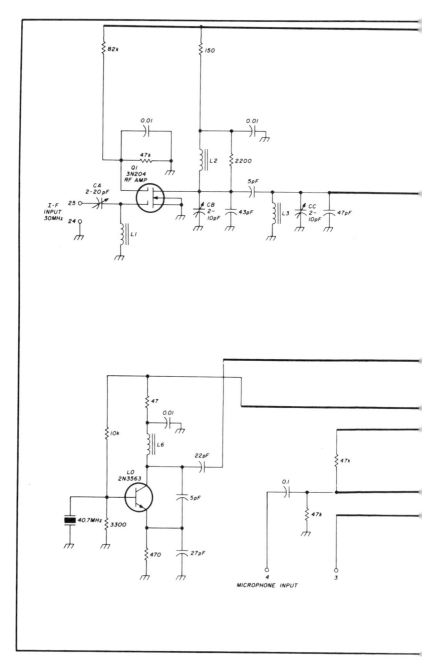

Fig. 6-3. Circuit diagram of the Advanced Receiver Research 30-MHz transmit/receive i-f unit for use with the 10-GHz Gunnplexer. See Table 6-1 for notes.

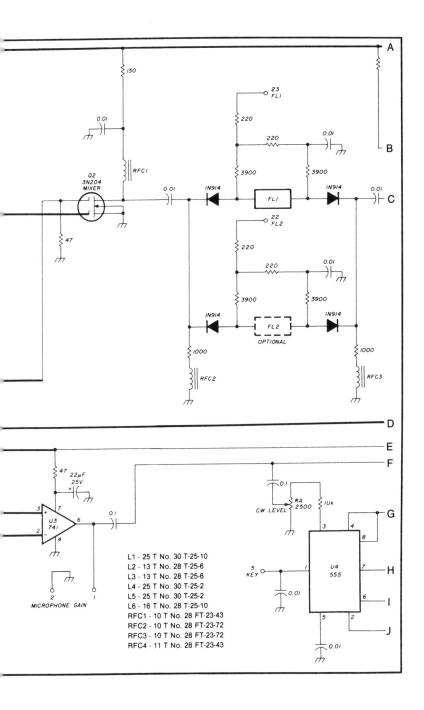

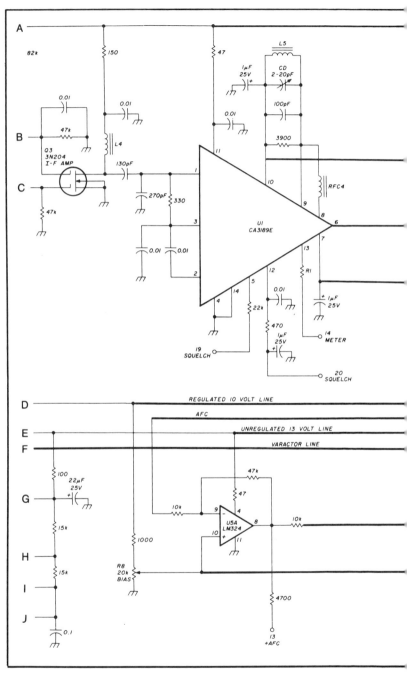

Fig. 6-3. (Continued from page 70.)

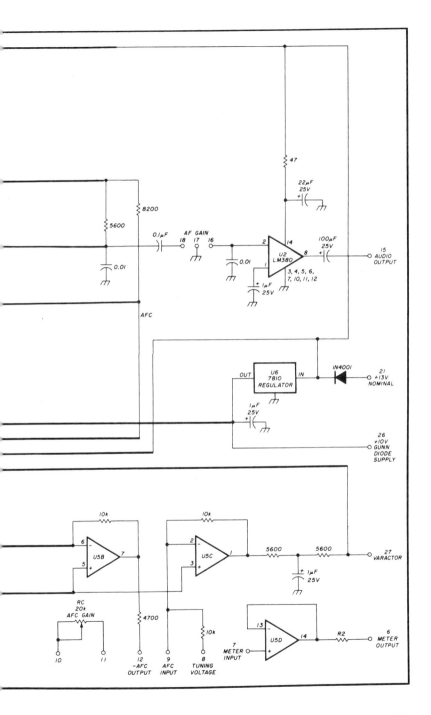

Table 6-2. Specifications of the Advanced Receiver Research 30-MHz Transmit/Receive i-f Unit in Fig. 6-3.

```
Bandwidth at 30 MHz..................................2 MHz
Bandwidth at 10.7 MHz
   at −3 dB..........................................220 kHz
   at −50 dB.........................................700 kHz
Sensitivity
   for 20 dB quieting..........................less than 0.2 uV
Image rejection.............................greater than 60 dB
I-f rejection...............................greater than 50 dB
Spurious rejection..........................greater than 50 dB
Audio output...................2.5 watts min. into an 8-ohm load
Supply voltage.....................13 volts nominal (12-16 volts)
Supply current ............250 mA (includes Gunnplexer current)
Size..........................................3 × 6 inches
```

volts at 250 milliamperes. The units are designed for operation over a temperature range of −25 to +65 degrees C.

The A.R.R. RXMR30VD circuit diagram is shown in Fig. 6-3, and an interconnection guide is illustrated in Fig. 6-4. Signals arriving at the i-f input (pins 24 and 25) are directed to a low-noise rf amplifier, Q1. A bandpass filter with a 3-dB bandwidth of 2 MHz is located between the rf amplifier and mixer (Q2). Local-oscillator injection for the mixer is provided by a crystal-controlled 40.7-MHz oscillator stage, while output from the mixer (10.7 MHz) is directed to either of two diode-switched ceramic filters. Voltage applied to pin 23 switches on FL1 (supplied with unit), and voltage applied to pin 22 switches on FL2 (optional). The output from this filter is directed to the i-f amplifier stage, Q3. The resultant signal is then routed to pin 1 of U1, the FM subsystem chip. Detected audio then moves through a level control (pins 16, 17, and 18) to pin 2 of U2, the audio-output amplifier. This stage has sufficient power to drive earphones and/or a speaker. Its output is available at pin 15 of the ARR circuit board. Squelch control is available at pins 19 and 20. Pin 14 may be connected to a signal-strength meter. See Fig. 6-4 for further clarification. Afc output from U1 is amplified by U5A. U5B inverts the amplified afc signal so either polarity afc may be selected by S3, while overall afc gain is controlled by Rc. U5C sums afc information along with manual tuning voltage, and this composite signal is applied to the varactor diode terminal of the Gunnplexer. U5D functions as a meter driver for center tune (discriminator zero), and manual-tuning-voltage indication. The driver's input is selected by S1. U3 is a microphone amplifier which boosts the audio to a level suitable for modulating the Gunnplexer.

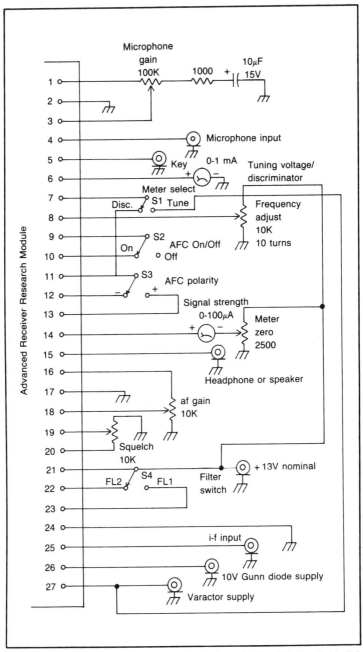

Fig. 6-4. Interconnection guide to the i-f unit shown in Fig. 6-3. See Table 6-2 for operating specifications.

U4 generates an approximate 500-Hz tone for modulated-CW (MCW) operation. Audio or MCW information is applied to the varactor diode of the Gunnplexer, along with the afc/tuning voltage. U6 is a three-terminal regulator to provide a stable 10-volt supply to the Gunn diode. A 1N4001 diode is included to provide reverse-polarity protection.

The RXMR30VD is factory aligned and doesn't require further adjustment. Should further adjustments be desired, however, the following procedure should be followed.

1. Connect a signal generator capable of delivering a 30-MHz signal to the i-f input connection.
2. Adjust the output level of the generator to a point where the RXMR30VD signal-strength meter just begins to move up scale.
3. Adjust C_a, C_b and C_c for maximum indication on the meter.
4. Temporarily disconnect the signal generator from the RXMR30VD.
5. Place S1 in the discriminator position.
6. While switching S3 between the + and − afc positions, adjust R_b so that the same meter reading is obtained in both switch positions. This reading should be mid scale on the discriminator meter.
7. Reconnect the signal generator to the RXMR30VD and center the signal in the receiver passband.
8. While switching S1 between the afc on and off positions, adjust C_d so that the same meter reading is obtained for both positions. This reading should be mid scale on the discriminator meter.
9. Adjust R_c for the desired afc gain/locking range by shifting the signal generator frequency away from the passband center.
10. R_c is adjusted while in communication with another station. Advance the level of this control to the point where distortion occurs. Back off the setting of the control to the point where distortion is no longer present.

Because the i-f/receiver stage is quite sensitive, the complete RXMR30VD should be mounted in a well-shielded enclosure. The Gunnplexer unit can be mounted close to the RXMR30VD, or it can be mounted on a tower and connected to the RXMR30VD through three lengths of coaxial cable. Although the Gunn diode and varactor tuning lines carry only dc and low-level audio signals, shielded cable is a must because even small voltage pickups can produce relatively large modulating voltages (which are undesirable). In cases of severe pickup, small rf chokes may be required directly at the Gunnplexer Gunn-diode and varactor connections of the Gunnplexer. Lengths of cable of up to several hundred feet between the Gunnplexer and the RXMR30VD have been used with no problems. For long i-f connection runs, it is best to install a 30-MHz preamplifier at the Gunnplexer to prevent the long cable run from adversely affecting the system noise figure.

Our special thanks to Advance Receiver Research of Burlington, Connecticut for sharing the previous information on their outstanding units.

Basic operations of this setup follow procedures outlined under the previous Quick and Easy 10-GHz system. Set one unit on a specific frequency and tune the other unit while using a 2-meter liaison until communications are established. Once that initial point of reference has been achieved, the involved amateurs can almost throw away their "vhf crutches" and proceed completely with 10-GHz communications and expansions.

A PHASE-LOCKED 10 GHz SETUP
FOR LONG-DISTANCE COMMUNICATIONS

This setup is what may be considered "top dog" in 10-GHz operations. It was originally constructed by Jim Hagan, WA4GHK, of Palm Bay, Florida, and used to establish high-quality voice communications over paths exceeding 80 miles. Jim is continuing to use and prove this system's capabilities, so it's quite possible that additional DX records will be achieved by the time this book appears in print. Since we feel WA4GHK should receive full credit for his system layout and its operation, his personal description and discussion follows. Our special thanks to WA4GHK for sharing this information, and we wish him total success in his future endeavors.

"Several years ago, I purchased a pair of the Microwave Associates Gunnplexers for the 10-GHz band. Previous experiments on 432 and 1296 MHz many years ago convinced me of the necessity of using very narrow band systems for serious long distance work with weak signals. I therefore decided that the Gunnplexer would somehow need to be frequency stabilized rather than merely operated as a free-running modulated oscillator as most hams use it. The best and most practical way of doing this appeared to be use of phase-locking techniques.

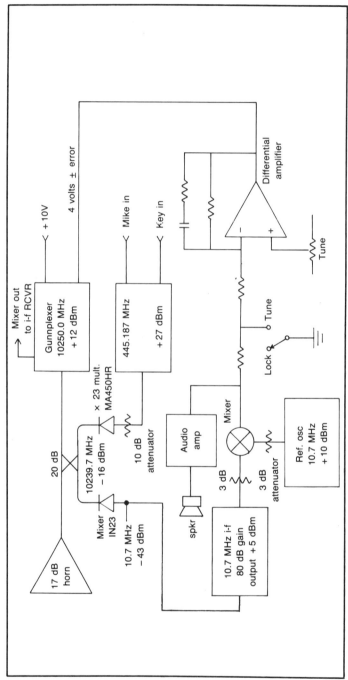

Fig. 6-5. The WA4GHK phase-locked Gunnplexer diagram.

"I spent the next year developing a practical circuit to do this using readily available components. A block diagram of the resulting system is shown in Fig. 6-5. A 440-MHz FM rig which has been modified for very narrow deviation (about 200 Hz peak) generates a modulated carrier at 445.187 MHz. This is multiplied 23 times to 10239.7 MHz by a crystal diode mounted in a section of waveguide. A sample of the Gunnplexer output (to be locked at 10250.0 MHz) is coupled into a second diode, and mixed with the multiplied signal by means of a cross-guide directional coupler. The difference frequency, 10.7 MHz is fed to an i-f amplifier that has 80 dB of gain. This amplifier is saturated and thus operates as a limiter. The output of the amplifier is fed to one port of a phase detector. The other port is driven at 10.7 MHz by a crystal-controlled reference oscillator. The output of the phase detector containing phase error information proceeds through a loop-error amplifier to the varactor-tune input on the Gunnplexer. An audio amplifier is provided to aid in getting the circuit locked initially. To lock the unit, the tune/lock switch is placed in the tune position, and the tune control is adjusted so that a rather rough and warbly beat note is heard in the audio amplifier. This is an indication that the Gunnplexer is within 10-15 kHz of 10250.0 MHz, or close enough for the phase lock circuitry to operate. At this point, the tune/lock switch is moved to the lock position. The Gunnplexer is then operating in the stabilized mode at exactly 10250.0 MHz, $\pm$ any errors in the crystal oscillator frequencies. A second Gunnplexer is similarly locked at 10279.9 MHz. A communications receiver capable of NBFM and CW operation, and a 30-MHz preamplifier are connected to the Gunnplexer mixer output. Narrow-band FM with 5-KHz deviation, or CW with a 500-Hz bandwidth, can be used. CW operation is accomplished by shifting the Gunnplexer off frequency by 2 kHz when the key is up. This is done because normal carrier on-off keying would cause the system to lose phase lock.

"In my system, a 4-foot and a 30-inch parabolic dish are used. Each Gunnplexer feeds its dish through a circular waveguide feedhorn made from common bathroom-plumbing hardware. A choke around the rim of the feedhorn ensures identical radiation patterns in both the E and H planes for optimum illumination of dishes in the 0.3 to 0.4 focal-length/diameter range. Feedhorn VSWR is tuned out by means of a 3-screw tuner built into a section of waveguide immediately behind the horn. The dishes are mounted on 3-foot Radio Shack TV-antenna tripods. See Figs. 6-6 through 6-11.

Fig. 6-6. Jim Hagan, WA4GHK, with 10-GHz setup ready for communications over a 25-mile over-water path. Jim is sighting the path with a compass and peephole.

Fig. 6-7. Close-up view of the electronics at the feedhorn of the 48-inch WA4GHK dish. Note choke ring around end of feed waveguide pointed at dish.

Fig. 6-8. The electronics of Fig. 6-7 seen from the opposite side. The i-f preamplifier is mounted atop the Gunnplexer. The Kenwood TS700A transceiver is the i-f receiver/transmitter.

"Local tests out to about 10 miles over obstructed paths seems to indicate that 10 GHz can provide about the same communications coverage as experienced on 2 meters with low power and ground plane antennas at low heights. The important thing involves trying to get both dishes above all nearby surrounding objects, whether or not a line-of-sight path exists between stations.

"Several long-haul tests over water have been carried out. CW communications were easily achieved at 50 miles even though both stations were only about 6 feet above the water. A late afternoon duct permitted excellent voice communications at 83 miles with signals better than 50 dB above the noise. Both stations were still only a few feet above the water. Two-meter FM gear carried along for liaison would not work across this path. Ducts of this type have been noted on two other occasions in my brief time on 10 GHz. I believe that over-water ducts exist fairly often, if not daily. They must obviously be very close to the water and only a few feet thick, as a 2-meter signal will not propagate through them. I believe these ducts have not been previously reported, because in normal microwave operations the antennas are mounted as high as possible and would probably be above these thin ducts. I intend to investigate longer over-water paths out to 500 to 700 miles and overland paths of 70-80 miles in the near future.

It is very important to be able to precisely aim both antennas for long over-the-horizon paths. A small hole in the back of the dish,

Fig. 6-9. WA4GHK sets the phase lock for a 10-GHz, 25-mile, over-water path. The signal will pass only a few feet above the water.

Fig. 6-10. A right-side view of a WA4GHK 10-GHz setup that uses a 440-MHz source, i-f amplifier, a reference oscillator, and loop amplifier.

and a wire pointer mounted on the feedhorn, are used like a gunsight to aim the dish at a predetermined point on the horizon. A surveyor's compass and an aircraft sectional chart are invaluable aids. A pointing accuracy of $\pm$ 1 percent is needed. Also note that on very long paths the actual azimuth will not be the same as measured from maps because of distortion caused in the projec-

Fig. 6-11. Left-side view of the setup in Fig. 6-10, showing the interconnections.

tion used to print the map. See '*Aiming Microwave Antennas*' in '*The New Frontier,*' June, 1981 *QST*, page 60."

The collection of Gunnplexer systems presented in this chapter represent a cross section of ideas. I trust you'll find them useful for starting and expanding your activities on this unique amateur band.

Chapter 7

Microwave Networking and Data Packeting

The use of expandable microwave-communications networks for interconnecting local, national, and international groups will open a new era of amateur activity during the 1980s. Boasting the ability to handle several modes of different bandwidths, such full networks can become the backbone of amateur audio, video, and computer interlinks. Through these arrangements, electronic mail boxing and related high-data-rate communications can also be implemented. Similar independent networks have already been started by radio amateurs in the United States and Canada (packet-radio networks). These networks are predicted to expand and possibly evolve into a nationwide system. At that time, additional small networks will grow until a sufficient number of branches can support a connection to the master system. These arrangements are not really "far out"—they are presently in their infancy at various locations within the United States. Many amateurs, for example, are presently communicating both by audio and modem-linked computers. This arrangement, however, reflects limitations in the number of simultaneous users and in rate of data flow. Microwave links, conversely, can handle large numbers of such users.

COMPUTER COMMUNICATIONS

The possibility of high-speed computer communications via amateur links was a natural evolution; its foothold of acceptance became legal with the FCC's authorization of ASCII (American Standard Code for Information Interchange) transmissions on the amateur bands. The current data rates "top out" at 1200 baud—a rather low speed which will soon give way to 4800, 9600, and higher baud rates. These high speeds are not utilized on HF bands because their bandwidths are simply too broad for that spectrum.

The arrangements described thus far can apply to two different kinds of links: conventional voice communications, and computer communications. A custom-designed network for implementing either of these modes may be utilized, or a dual-purpose arrangement may be more useful. Computer-linking networks can rapidly become quite sophisticated in architecture, progressing far beyond the scope of this book. We thus suggest as additional reading *Packet Radio*, (TAB Book No. 1345). Additionally, the ARRL (American Radio Relay League) and AMRAD (Amateur Radio Research and Development Corporation) support several development programs in this area. A number of truly fascinating papers on existing computer networks (usually linked by 146 mHz or 440 mHz) and on projected networks, plus many revolutionary concepts were presented at AMRAD's first joint conference held on October 16 and 17, 1981, at the National Bureau of Standards, Gaithersburg, Maryland. Several noted Packet Radio pioneers attended this gathering, presenting a wide variety of papers which will surely set the pace for future activities. The subjects covered range from packeting protocols and handshakes, through design of headers and trailers in packets, to discussions of existing systems and design suggestions for larger networks. The network described on the following pages is one I presented. Additionally, the expandable network was also carried in the pages of *ham radio* magazine, August 1982. My special thanks to the editors of *ham radio* for permission to include the network's story herein.

AN EXPANDABLE NETWORK FOR MULTIMODE COMMUNICATIONS

The projected network described on the following pages was originally conceived by the author for the purpose of interlinking communities and cities on a broadband basis. Numerous other capabilities, however, were soon visualized and included to permit almost direct compatibility with future expansions. The resultant network became a highly flexible system which may be used between adjoining communities or cities, with additional networks being implemented in other areas, and interlinked as desired. Communication modes that can be handled by the network are limited only by user's desires and modes. An outline of the microwave network is shown in Fig. 7-1, and an overview of its operation follow.

The primary purpose of the microwave network is to provide emergency communications between areas or cities normally separated by a distance greater than their normal 2-meter communications range. Secondary communications capabilities should be considered at installation time, however, since path losses and overall network bandwidth are directly related. The number of "passive" microwave repeaters will be determined by distance and terrain between associated cities, each city accepting responsibility for their part of the link. Existing 2-meter repeater groups and councils can provide finances and frequency coordination. Two transceivers are shown connected to each microwave port: one preset to the primary frequency, and the other scanning an approximate 1-MHz range of the 2-meter band (Exception: all secondary transceivers have primary-frequency lockout). Secondary transceivers are under microprocessor control, permitting frequency scanning, spread-spectrum operation, tone control of transceiver functions (enable/disable, lockout, connect to mailbox, etc.). The network could initially develop between any two areas (each preferably with at least two local 2 meter repeaters, since this would confine the costs of microwave link additions). Additional areas could join the existing network by financing their section users. A similar network may be installed in another area, systems may grow until an overall network merger is warranted. Additional networks may, likewise, grow and merge with the existing system as desired. Further expansions may include branches and subnetworks as desired.

Continuing the network a step further, interlinking with the OSCAR Phase IV geostationary satellites could provide expanded coverage for compatible modes (projected date 1986). The outline for this concept is shown in Fig. 7-2. OSCAR Phase IV is slated to include several concepts applicable to data communications. Some of these features are dedicated channels, tone controlling, and mailboxing. In some instances, a microwave network port may interface with an OSCAR earthbased transponder. Other times, a separate network-to-satellite earth-based station will be required. The criteria will, naturally, be determined by geographic locations of microwave links.

OSCAR satellites necessarily utilize narrowband modes such

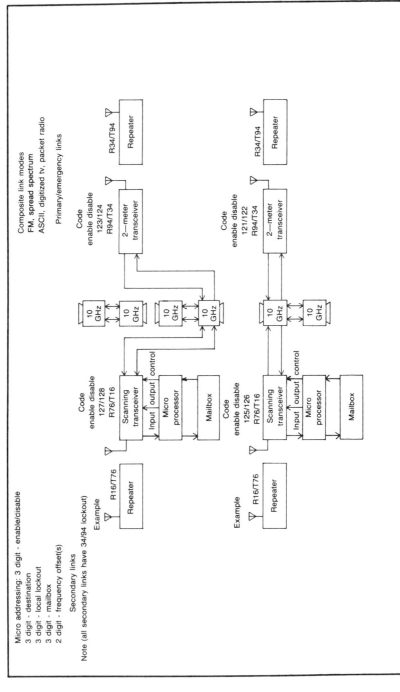

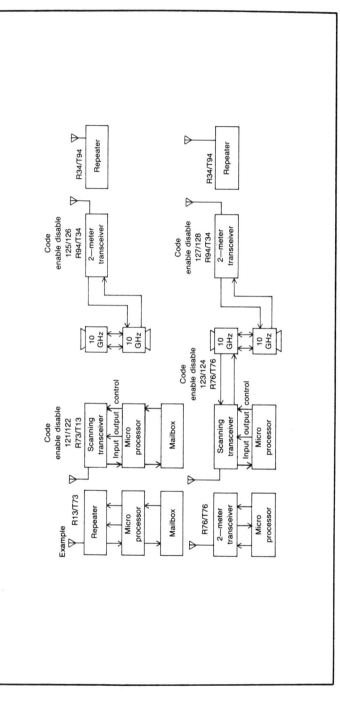

Fig. 7-1. Suggested national microwave network for providing emergency communications between areas separated beyond normal 2-meter coverage (courtesy *ham radio* magazine).

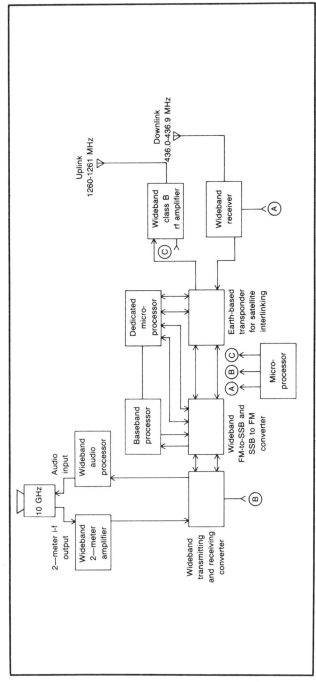

Fig. 7-2. OSCAR Phase-IV interlink for a national microwave network. Compatibility is ascribed to PLL-SSB concepts. A master microprocessor monitors the control frequency of 2 meters and the activity of the 2-meter passband to enable the satellite link of a preestablished number of signals on the downlink, uplink, and 2-meter passband is not exceeded. If justified, a proper follow-up sequence of 2-meter control signals will access the satellite link (courtesy *ham radio* magazine).

as SSB or CW; however, a microwave network should utilize a constant carrier mode such as FM. The key to compatibility between these modes is constant-amplitude single sideband, or merely PLL-SSB. This concept, which was developed in Europe 5 or 6 years ago, employs variable amplitude in the normally suppressed carrier. Carrier amplitude is small during modulation, but increases to full power during breaks of speech (after passing through the microwave network, the carrier may be fully removed, resulting in conventional SSB). Finally, microprocessor control is used for the link, its preprogrammed functions being available for call-up coded tones.

The concepts associated with microwave links are, in several respects, unlike those employed in conventional VHF repeater links. Bandwidths of microwave systems, for example, are typically 0.5 to 4 MHz. Outer-power levels are noticeably lower, with large parabolic dishes providing signal gain. Conventional superheterodyne techniques are also altered: each microwave transmitter operates continuously, with a small portion of its output power being directed to its receiver's front end to provide a local oscillator signal. Transmit frequencies of communicating units are then offset by a difference equal to the desired i-f (as was discussed in previous chapters). This arrangement may be visualized with the aid of Fig. 7-3. All microwave units are originally transmitting on their hypothetical resting frequency. An incoming signal on 146.00 MHz shifts the transmitting unit 146 MHz (A second signal on 146.50 and a third signal on 146.80 would appear as subcarriers of the original signal, until the 146.00 disappeared. The 146.80 signal would then be a subcarrier of the 146.50 signal.) Assuming a relay is required between ports, it would receive the 2.146-GHz signal, heterodynes to 0.146 GHz, amplify the signal(s) and apply it to the associated transmitter. The 2.146-GHz signal would then be received at the subsequent microwave port, converted to 146 MHz and applied to a broadband amplifier. That amplifier's output would feed the next 2.1-GHz transmitter and the 146-MHz transceiver. The microwave link's overall bandwidth could easily expand to 1 MHz, as necessary. All operations and frequencies of network 2-meter transceivers are under microprocessor control. This means that signals may be selected or rejected by tone control, as desired. Preprogramming of the microprocessor establishes the basic network standards.

Two microwave bands are prime candidates for network links: 2.3 GHz and 10 GHz. Gunnplexers are readily available for 10-GHz

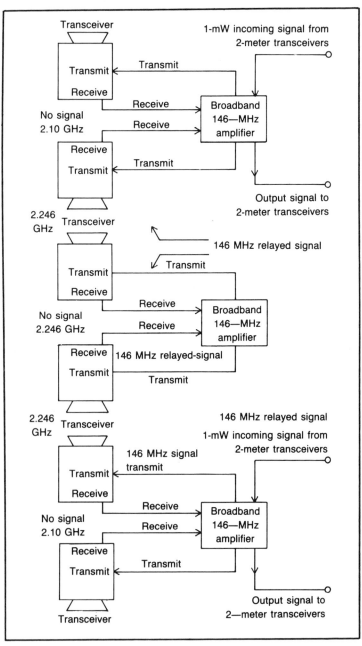

Fig. 7-3. Operational concept of the national microwave network. Middle unit is considered a "dumb," or passive, repeater with intelligent ports located in large cities (courtesy *ham radio* magazine).

systems, however their range is limited. Similar units for operation on 2.3 GHz will soon be available from Universal Communications, P.O. Box 339, Arlington, Texas 76010. The 2.1-GHz units provide output of 100 mW or 1 watt, as required. Cost of 10-GHz Gunnplexers is approximately $120 each. The 2.3-Ghz units are approximately $180 each.

Use Fig. 7-1 as a reference during the following brief technical discussion.

Assume an amateur operating on 146.76 MHz desires to contact a distant repeater on 146.76 MHz. A signal with tone-coded squelch and tones of 1, 2, and 3 are used for connecting the scanning transceiver into the network. Notice the distant transceiver employs lockout, preventing accidental access. Another 3 digit code (1, 2, 7) brings up the desired distant scanning transceiver, with subsequent microprocessor control establishing operating parameters for accessing that area's 146.76-MHz repeater. Assuming the distant amateurs desire disconnection from the link (or the calling station desires distant disconnection) another three-digit code will shut off that transceiver (example: 128). Data packets may be moved either to the distant repeater, or left in the electronic mailbox as required. Continuing overall system capabilities one step further, we can use tone control and portlocated microprocessors for handling frequency offsets and Spread Spectrum hopping sequences. This capability would permit an individual amateur operating on 146.52 MHz to catch the network's scanning transceiver, establish different network frequencies, and proceed as described. A full description of the network would, obviously, encompass numerous pages of discussion. We thus leave those operations open for your imagination and thoughts of expansion. The network outlines is a coarse system for future communications techniques. We hope this first step will inspire future developments.

PACKET COMMUNICATIONS

The mechanics of data-packet communications resemble a form of electronic mail service. Any station connected to the associated network has the ability to originate, transmit, and receive packets; each specific address being determined by a particular protocol. Each user's terminal is thus setup to respond only to its designated call. A standby electronic mailbox can assist communications when the operator is absent.

Two popular forms of packet communications are in use today. First is an arrangement whereby any station in the group may generate messages for another station and transmit them anytime the channel is clear. This technique has been shown to result in occasional packet collisions, requiring terminals at each end to recommunicate the message until a perfect copy is acknowledged. The second arrangement involves a master, or "host" computer which polls all stations on a round-robin basis, asking those stations if there is a message to be moved into the network. If there is a message, it is either moved to the specified receiving station or held in an electronic mailbox at the host computer until that station's terminal responds to polling.

Both of these techniques have been used by packet pioneers around the United States. Research continues in this area, and a specific determination as to the "best" arrangement has not been concluded. Both arrangements are quite promising and either could be successfully used in future systems. The utilization of microwaves for packet communications allows any combinations of data formatting and system operations, either independently or simultaneously, plus several audio channels may also be included in that operation if desired. The expandable microwave network mentioned in this chapter is capable of handling either packeting format.

For more information on packet radio, see *Packet Radio,* (TAB Book No. 1345).

Chapter 8

Power Supplies for Microwave Systems

Although power supplies for microwave equipment are not really all that different from those used for other equipment, locating a suitable schematic when needed can occasionally become a challenge. While basic power supplies are relatively easy to construct in little time, finding the exact transistor pin-out, diode polarity, etc., can present distractions. Therefore, several inexpensive and easy-to-assemble units particularly suitable for projects in this book are presented in this chapter. I hope this information will serve as a convenience for thus rigging a microwave system of your choice. Before delving into the selected circuits, however, let's briefly review some possibly overlooked considerations in junkbox and weekend construction practice.

TRANSFORMERS

Many "junkbox" power transformers can be adapted for use in home-constructed power supplies. Output voltage can be measured with a meter, but measuring current capability is a different matter. As a rough means of current calculation, the old "finger" method may prove useful. Simply stated, placing one's fingers parallel to a transformer's center laminations and estimating 100 milliamperes per finger width will provide a reasonable current calculation. As an example, old TV transformers are usually three or four fingers wide in laminations. This translates roughly to 300 or 400 milliamperes current capability. Small transformers, with their "one-finger-width" laminations, are probably capable of 100 milliamperes current, etc. Although not a laboratory-type calculation, this technique is usually acceptable for most amateur applications.

CAPACITORS

Large values of capacitance are quite desirable and necessary for low-voltage (solid-state) power supplies. Filter capacitors serve as ac resistance according to their capacitive reactance. A 40-μF may act as a 10-ohm resistor to ac, for example, causing a resultant 1 volt drop. Assuming this capacitance is incorporated in a 300-volt tube-type power supply, the 1 volt ac to 300 dc ratio would be negligible and its output ripple would be quite acceptable. If this same 40-μF capacitor was used in a solid-state (low voltage) supply, its ac resistance (capacitive reactance) would reflect the same 10 ohms resistance and 1 volt drop. A 12-volt dc output power supply would thus exhibit 1 volt ac ripple; a 12 to 1 ratio which, naturally, is intolerable. The solution involves using large values of filter capacitance, producing substantially greater capacitive reactance (ac resistance). Assuming, for example, the 40-μF capacitor is replaced with a 1000-μF capacitor, approximately .001 volt ac would be dropped across this capacitive reactance; a quite acceptable level for low voltage supplies (9 to 18 volts). The previous discussion is not to be taken as a killjoy. Simply remember that adding large values of capacitance to home-constructed power supplies is advantageous. Power supplies in the 9 to 18 volt range operate very well when employing filter capacitors between 1,000 μF and 5,000 μF. Those "hamfest bargain" filter capacitors of many thousands of microfarads are additional power supply components which definitely shouldn't be overlooked in this quest. Placing a couple of these monsters in a power supply ensures maximum current-producing ability and pure dc output. Output filter capacitors and their load (circuit in use) form an R/C time constant: because the load's internal resistance is a fixed value, increasing the capacitance increases the R/C constant. When the capacitor has a longer period to deliver stored energy, the energy is not depleted and higher output currents are possible.

REGULATORS

Solid-state regulators are fantastic little devices, but they are

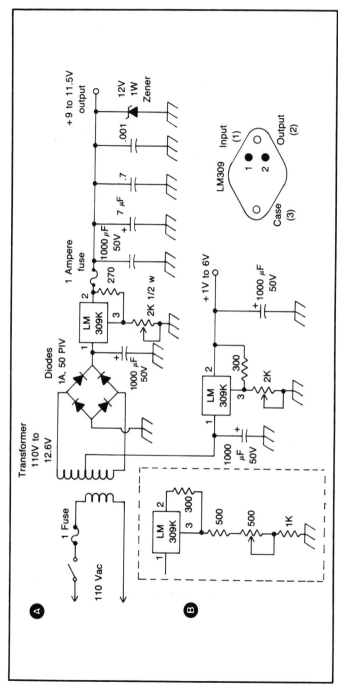

Fig. 8-1. A rugged, general-purpose power supply. The inset (B) shows an operational voltage-span-limiting circuit that can be used with this, or another supply.

not indestructible. Play it safe, and bolt a heat sink to their case. This heat sink can be relatively small for low current supplies; it can be cut from a small piece of aluminum or copper. Additionally, be sure the selected regulator is capable of handling the required current over extended periods of use. Feel the heat sink during operation: you should be able to keep your finger on it for several minutes. The popular LM309 regulator is available in a number of packages (case styles). Each particular package is capable of different maximum current. Choose according to needs; if in doubt, simply choose the largest available package—and use a large heat sink.

Keep the previous tips in mind and few problems should be encountered in overall system performance. The result of these efforts will be repaid tenfold in smooth equipment performance.

A RUGGED GENERAL-PURPOSE POWER SUPPLY

The power supply shown in Fig. 8-1 is straightforward in design, yet a perfect match for most of the microwave projects presented in this book. The output voltage may be varied from approximately 6 to 12.5 volts. Current capability is determined by transformer size and the LM309 regulator rating. Assuming the unit is set for 11 volts output for a Gunnplexer, a one ampere LM309 is suggested for long-term reliability and stability. If the supply is used for a 2.3-GHz system, the voltage-adjust potentiometer can be calibrated for tuning the desired i-f range. Should additional tuning be desired, however, the feedback resistor between pins two and three of the LM309 can be varied ± 20 percent of the value shown. Additional vernier tuning over the selected voltage range can be accomplished by the optional circuit of Fig. 8-1 B. This range-limiting circuit, incidentally, may be applied to any power supply using a single variable resistor for adjusting voltage output. Voltage drops are merely spread among three resistors rather than one resistor. Values can be swapped to obtain the desired level; merely remember to keep the full string equal to the replaced-resistance value.

An optional second regulated output voltage may be obtained from the transformer's center tap, if desired. This trick may also be used with most popular power supply circuits. Notice the second LM309 regulator has a separate potentiometer for independent voltage adjustment.

The capacitor values shown are approximate; values between 500 and 75,000 μF may be used. The bank of capacitors on the output eliminate spurious power-supply coupled oscillations. Each capacitor bypasses a different range of frequencies, with the overall results providing bypassing of the full rf spectrum. An optional Zener diode and fuse are shown as a protective device. This circuit may also be employed in other power supplies of similar design. If the output voltage rises above 12 volts, the diode will conduct, the fuse will pop and protect the circuit. Remember to check the circuit carefully and set the output voltage before connecting the load. Readjust the output voltage as required after load connection. The result should then be smooth and proper operation.

SAFE-STOP POWER SUPPLY

This power supply, designated according to its self-protect

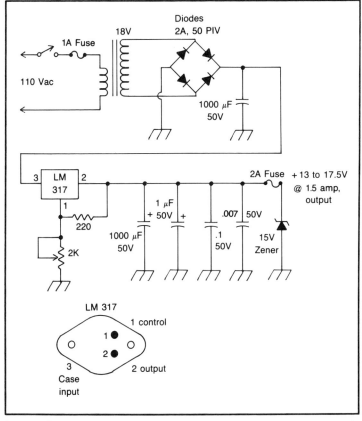

Fig. 8-2. Safe-stop supply features overcurrent and thermal protection. The 18-volt zener diode and the 2-ampere fuse are optional.

ability, shuts off output when excessive current demands (over 1.5 amperes) or dead shorts are experienced. See Fig. 8-2. The high current LM317 provides a stable output under varying loads. If load current becomes excessive, the LM317 will go into thermal shutdown until the overload is removed. After the LM317 cools, operation returns to normal and output voltage is restored (such features should be incorporated in all power supplies).

Circuit design of this supply is similar to the previously outlined general-purpose unit, and may use similar junkbox parts. When selecting a transformer, bear in mind at least one-half volt (and no more than 5 volts) should be dropped across the regulator for best overall performance. Likewise, a heat sink of appropriate size should be employed with the LM317 if the device isn't mounted directly to a metal case (remember to use insulating mica washers and heat-sink compound).

The filter capacitors, as previously outlined, may typically be any value 1,000 μF or higher. Also, the voltage-limiting arrangement illustrated in Fig. 8-1 may be included in this supply if desired. Output voltage is adjusted before connection of the load circuit, and readjusted to compensate for current demands after circuit connection. As in the previous supply, the bank of capacitors located on the regulator's output prevents power-supply-coupled oscillations which might occur at various frequencies. The fuse and Zener diode are also optional, affording a high degree of overvoltage and transient protection.

The circuit can be constructed on perforated board, or the larger parts can be mounted directly to a metal cabinet with solder lugs being used for holding smaller components such as diodes, resistors, etc. Either arrangement seems to work well; the prime consideration being a neat layout with regulator "breathing room." The LM317 is a rugged, high-quality regulator that should perform well in this power supply. Currents up to 1.5 amperes may be drawn through this device; a more-than-ample figure for usual microwave setups.

THE PICK-A-VOLT SUPPLY

The supply shown in Fig. 8-3 is relatively conventional in design except for the diode voltage-dropping string on its output. This arrangement is based on the fact that silicon diodes such as those used in power supplies produce a constant 0.7 volt drop between their anode and cathode. As a result, various voltages can be ac-

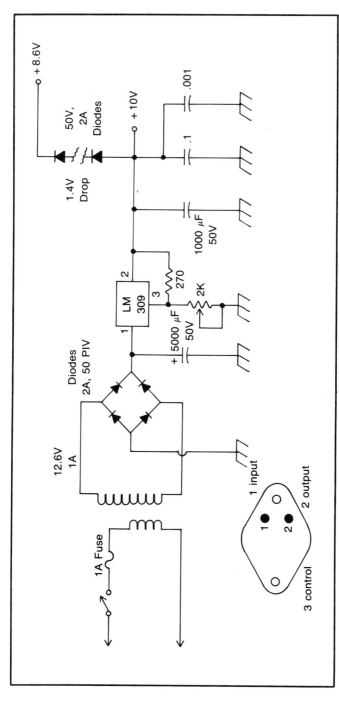

Fig. 8-3. The pick-a-volt supply. This unit operates on the principle that each silicon diode provides a 0.7-volt drop. The diodes can be stacked to provide any amount of drop desired.

quired by stacking the necessary number of diodes in series with the power supply output and load.

The benefits of using series diodes for second voltage outputs include obtaining levels between full and half transformer voltages, and an additional amount of regulation at the second voltage output. Because the diodes are functioning in their forward-biased region, this constant voltage drop is used to maximum advantage. The only restraint to using such diode strings is ensuring all diodes are similar types and capable of continuously passing the current drawn through them. With the many bargain packages of diodes available, one can often save several dollars when using this technique. Note that regulator adjustment affects both voltage outputs. This capability may be desirable when powering several Gunnplexer systems from one supply.

NICKEL-CADMIUM BATTERIES

The use of nickel-cadmium batteries to power microwave circuits and systems is a logical consideration. Since a respectable amount of weak-signal experimentation and line-of-sight DXing is conducted in this range, portability is an often-considered situation.

Nickel-cadmium batteries are available in numerous shapes, styles and current capabilities. A 10-GHz system that usually operates in the 10- to 25-milliwatt range can easily be powered by AA-size 450-milliampere-hour cells arranged in a 9 to 10.5 volt pack. If extended periods (over 10 hours) of continuous operation are contemplated, the AA cells may be replaced with C cells. Several of the popular 2-meter FM talkie battery packs (9.6 volt types) are suitable for powering Gunnplexers; however, battery packs with outputs near 11 volts should be avoided: their full-charge output often exceeds the Gunnplexer's 11-volt maximum rating.

Trickle-charge, and large storage-battery arrangements are often beneficial for remote microwave setups. The battery acts as a large filter capacitor on the supply's output until ac power is removed. The microwave setup then draws current from the batteries until ac power is restored.

NATURAL POWER SOURCES

Microwave systems and natural-power sources are complementary. The capability of powering remote stations in a self-sufficient manner opens a number of unique opportunities. The most common methods of acquiring natural energy involves solar cells, wind generators, and hydroelectric power generators. Although limited amounts of energy are available from sample basic systems, the power requirements of most microwave units are relatively low. Construction of wind- or water-power setups usually center around the use of surplus or discarded automobile alternators, voltage regulators, and lead-acid storage batteries. Gear or belt-drive systems provide higher shaft speed, and increase efficiency to respectable levels. Energy is drawn directly from the natural power source as available, with the batteries providing excess storage, and backup during periods when natural energy is not available. Solar-power systems use stacked arrays of silicon photovoltaic cells for converting sun energy to power that maintains a charge in storage batteries.

Water Power

The basic outline of a hydro power system is shown in Fig. 8-4. Water moving through a stream, or falling from a drain ledge after a rain, causes a paddlewheel to rotate. This movement is "amplified" by a pulley arrangement, causing the auto generator to spin rapidly and charge its associated storage batteries. Energy is then drawn from the batteries as required. The full voltage-regulator setup salvaged from a wrecked auto is used; this provides a system identical to automotive battery charging, complete with ammeter. Currents are calculated according to electronic equipment used, storage-battery capacity, and generator output. As an example, assume the total current drawn by a microwave system is 250 mA (from a 12-volt system). Ten hours continuous operation would thus require 10×0.250, or 2.5 amperes. Since a general purpose auto lead-acid battery usually stores 150 ampere-hours, approximately 600 hours operation is possible between charges (150 A $\div$ 0.250 A = 600 hours). During this time, however, the battery is intermittently being recharged. The results are a constantly-powered microwave system of high reliability.

Wind Power Systems

A few years ago, the trend of using wind power to produce moderate amounts of energy became quite popular in the United States. Using turbine-type arrangements mounted atop variable direction mast or towers, these air current-driven systems supplied recharging capabilities to a small bank of lead-acid storage batteries. The complete arrangement provided sufficient power for operating

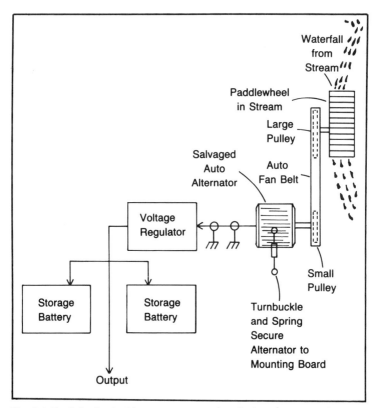

Fig. 8-4. Basic hydroelectric power system. Auto junkyard parts can be used to provide energy for a 12-volt system.

a typical 200-watt SSB transceiver approximately 1.5 hours each evening. The wind-power systems were particularly attractive to amateurs living near coastal areas or atop mountains. Since these areas experience almost continuous windflow, harnessing of that energy proved a quite logical decision. See Fig. 8-5.

Although popularity of wind generators has declined slightly, their use for natural energy definitely should not be overlooked. Indeed, a combination of both wind and water power is an ideal "final touch" for solar power setups. As a means of determining available energy force winds, one may run his own measurements with an anemometer or query a local weather bureau. Bear in mind energy force rather than mere prevalent winds are used for wind turbines. Assuming one is located in a sufficiently windy area, he can construct his own unit from scratch, or alternately, purchase one of the wind generator kits advertised in monthly publications.

Individual items, such as feathered or variable-pitch blades can also be purchased from kit suppliers. The home constructor may thus bypass critical tasks of designing aerodynamically efficient propeller systems. The optimum wind turbine should be capable of unobstructed movement in the horizontal plane. This is usually accomplished through the use of slip-ring connectors and one of two large blades on the turbine case's rear. The other large blade swings upward in an energy-generating wind, and drops horizontally during low-wind periods. This particular movement provides connection/disconnection of alternator and batteries to prevent still times discharge. An efficient wind turbine should also be capable of withstanding excessive winds without flying apart or blowing over. Finally, a voltage-regulator system is used with the storage batteries to prevent overcharging during periods of high wind.

Fig. 8-5. A modern, 1500-watt, continuous-duty wind-powered generator provided by Windpower Corp. of Livonia, Michigan. The unit charges a bank of 12-volt batteries located at ground level.

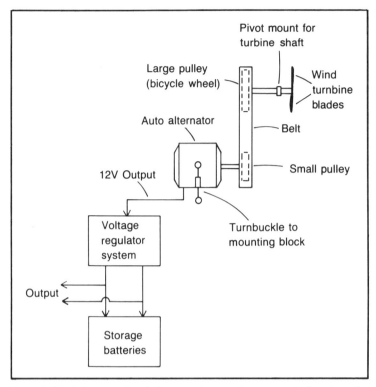

Fig. 8-6. Block diagram of a wind-powered generating system. Pulley size ratio provides increased speed necessary for efficient alternator performance.

The general layout of a basic wind power system is shown in Fig. 8-6. Either commercially obtained or full home-constructed propeller blades are attached to a central drive shaft. A large pulley (such as a bicycle wheel) is attached to the opposite end of this shaft. A low-friction shaft support and thrust-bearing assembly is used on the windshaft proper, while a small board and floor-flange setup allows horizontal movements. The generators/alternators can be reverse mounted to afford a high degree of rigidity while conserving space inside the turbine's cover. A single 12-volt alternator is sufficient for storage battery recharging; however, two properly phased alternators afford greater overall system reliability and output. An alternator can be placed on each side of the turbine's center wood mount, resulting in equal weight balance and smooth pulley operation. The complete turbine electrical system may be enclosed by a home-constructed cover; otherwise, the alternators will require weather protection and exposed moving parts will require monthly regreasing. Large two-conductor cable (connected through a slip-ring assembly on the floor flange, according to personal designs) then moves power to the ground unit.

The ground-located setup consists of a bank of 12-volt dc lead-acid storage batteries, a voltage regulator arrangement and a metering setup. The exact number of required storage batteries will depend on individual system applications. Most auto batteries are stamped with their associated ampere-hour rating. This figure can be followed, providing batteries are new. Otherwise, their ratings should be decreased 25 to 50 percent. Likewise, full-time use of discarded or rebuilt batteries should be avoided: they tend to freeze or discharge at the most inopportune time. The wind energy enthusiast should also study proper care of storage batteries to avoid premature failures and to ensure their long life.

The metering-setup, voltage-regulator system and alternator(s) may be obtained from a local junkyard, if desired. Assuming that the electrical system is secured intact from a single automobile, system interconnection should be quite simple. Chrysler Corporation electrical systems have proven to be quite useful and reliable. Additionally, separate panel-mount meters (Simpson, Triplett etc.) may be used in lieu of auto equivalents. Protective housing for the batteries and a small hydrometer for checking specific gravity should complete the list of essentials for system maintenance. Bear in mind, also, leads from batteries to load should be kept as short as possible; dc cannot be successfully transferred over long distances.

Because an amateur microwave setup requires little power, wind generator systems can usually be employed quite successfully. A typical microwave transceiver, for example, draws less than 700 milliamperes during receive and a maximum of two amperes during transmit. Assuming heavy use (and extended transmit time), we might thus calculate one ampere for each hour's on-the-air operation. Since a single lead-acid battery typically exhibits 150 ampere-hour storage capability, and since a battery preferably shouldn't be discharged over 50 percent, we can still realize at least 3 to 3.5 days operation during low wind times. A diminishing charge, however, will eventually gain ground on the low power system and cause battery charge to drop below that 50-percent level. This condition should occur only during periods of truly extensive use, such as emergencies. Otherwise, wind generator systems should prove quite useful for the typical amateur microwave QRP setup.

Solar Energy Systems

Renewed interest in solar-energy techniques has recently become quite apparent in the United States and, indeed, throughout the world. These power-harnessing techniques are not limited to merely electricity, but also include solar room heating, highly efficient solar water heaters, etc. Realizing an unlimited amount of energy from the sun reaches earth each day, utilization of this perpetual source is quite logical.

The general setup for a solar-powered electric generating system consists of one or more solar panels, a voltage regulator/charge-control unit, one or more high-quality storage batteries, and a metering setup for monitoring system operation. See Fig. 8-7. The number of solar cells and storage batteries will be determined by desired ampere-hour rating and wattage. As a beginning point, the popular Encon SX100 solar-cell arrays are capable of producing an approximate output of 14 volts at 2 amperes in direct sunlight. As mentioned earlier in this chapter, the typical microwave transceiver draws an average current of one ampere during a period of one hour's operation. Assuming use of that transceiver, along with an SX100 solar panel and a lead-acid battery capable of storing 100 or more amperes, the unit could be used almost nonstop. Pursuing this arrangement a few steps further, let's assume a low power transceiver is to be used in conjunction with the previously mentioned system approximately 4 hours a day: 1 ampere × 4 hours = 4 ampere-hours power required. The next day, a hypothetical 4 hours of sunlight would provide 4 (hours) × 2.2 (amperes), or 8.8 ampere-hours. Because the system employs a voltage regulator, however, recharging is complete without damage or overcharge. Assume further, a small solar power system consisting of a 4 ampere-hour or larger motorcycle battery and a small (50 mA) solar panel such as the inexpensive Encon number 5430 is used with the QRP transceiver. Approximately 2 hours of operation will draw 2 ampere-hours from the battery, leaving an approximate half charge. At least 40 hours of sunlight will be required for replacing that 2 amperes (.050 amperes × 40 hours = 2 amperes). Meanwhile, only 2 amperes remain in the battery until full recharging takes place. Size and capability of a particular solar-power installation will be directly determined by the amount of energy required and its particular hours of utilization. Bearing these

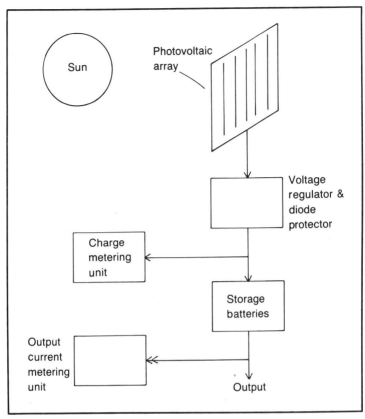

Fig. 8-7. A solar-energy system uses photovoltaic array to generate charging current for storage batteries. Popular systems similar to this design are appearing throughout the country. One manufacturer is ENCON Inc., of Livonia, Michigan.

thoughts in mind, we can now consider design of a solar power system for amateur use.

One typical solar-power electrical system is shown in Fig. 8-7. One or more SX100 solar panels are utilized, depending on one's particular current/power requirements, storage battery(s) capacity, etc. The SX100 panel itself consists of forty 10 by 10 cm solar cells connected in series. These cells are quite tolerant of heat buildups that can result from bright sunlight. Partial illumination can cause reverse biasing of cells and consequent heat increase. Cells of newer design, such as those used in the SX100, however, withstand such temperatures without adverse effects.

Output from the solar panel is directed via relatively large cable to the weather-protected charge-control and storage-battery system. This battery setup is similar to those described earlier in this chapter. One or more 150 ampere-hour lead-acid batteries are used in conjunction with a voltage control and metering arrangement. These items may be purchased from photovoltaic suppliers such as Encon Corporation, or salvaged from auto parts. Purchasing new items specifically designed for solar system use is, however, suggested.

Solar-power systems can be used quite effectively in most areas of the United States. The only criteria is an average amount of sunlight for daily operation and a sufficient storage capability to persevere on heavily clouded days. QRP systems, with their associated moderate power requirements, can operate over extended clouded or sunny days with such systems. Additional information on solar energy is available from ENCON Corp., 27584 Schoolcraft Rd., Livonia, Michigan, 48150.

Additional information on natural power systems can be found in *How to Make Home Electricity from Wind, Water, and Sunshine* (TAB Book No. 1128) and *Making and Using Electricity from the Sun* (TAB Book No. 1118).

Chapter 9

Setting Up, Tuning, and Operating Microwave Systems

The techniques of setting up a microwave system are relatively straightforward, but they should be carefully considered at the time of initial installation. This will permit future system links, bandwidth expansions, etc., to take place with minimum changes to the original setup. Environmental considerations are also making their marks in both amateur and commercial microwave areas. Tower and parabolic-dish zoning restrictions, plus concern for rf heating effects on body tissues, will remain a concern for amateurs in years to come. However, through careful planning and layout of a microwave system, larger repeating links can be situated in relatively clear remote areas, and accessed by small, low-power systems in any number of metropolitan areas. This arrangement is useful for multiple-operator systems. However, individual amateur activities are somewhat a different matter. One doesn't usually sit in his shack "low-band style," and operate 10-GHz by calling CQ, etc., particularly when two or three amateurs are trying their DX capabilities. Mobility and portability are the keynotes: one must travel to the high spots or make his way to suspected signal-ducting areas. Yes, signal ducts for frequencies of 10-GHz do exist—otherwise we wouldn't realize such 10-GHz DX paths as 300 or more miles. Returning to more typical lines, conventional 10-GHz Gunnplexers operating in conjunction with 2-foot parabolic dishes (as outlined in Chapter), and providing only 20 milliwatts of power, should be capable of communicating over distances of 200 to 250 miles. This assumes a nominal 200-kHz i-f bandwidth. Signal ducting is usually formed between temperature gradients in the atmosphere over water during the early morning hours. This phenomenon is occasionally detectable by the appearance of a gray or "dusty" cloud low on the horizon during early morning (or early morning fog over the water). Even without ducts, 10-GHz Gunnplexer communications should easily cover 30- to 50-mile ranges in metropolitan areas, provided line-of-sight paths are obtainable. Buildings, trees, and similar objects severely attenuate communications range. Flat and clear distances such as seaside areas provide reliable communications usually beyond 50 miles. A general survey of the area can produce a relatively dependable communications-range calculation; simply study the area for its "openness" while bearing in mind that 10 GHz exhibits characteristics quite similar to ordinary light.

CHARACTERISTICS OF 2.3 GHz AND LOWER

Other characteristics are found at lower frequencies in the microwave spectrum. At 2.3 GHz, for example, ranges are slightly longer than at 10 GHz, and signals are less effected by reflections. One might thus logically expect 2.3-GHz signals to pass through relatively mild rains and small amounts of foliage which would severely attenuate a 10-GHz signal. Parabolic dishes are also quite popular antennas for 2.3 GHz; the most commonly used items again being the snow scooters and Sno Coasters. Next in popularity are home-constructed parabolic reflectors 4-feet or more in diameter, cigar and slot-type antennas, and, finally, funneltype antennas. See Fig. 9-1.

Signal ducting at 2.3 GHz is also common, and follows lines similar to those experienced on 10 GHz. One of the unique aspects of signal ducting is its tendency to "run in layers." As an example, there may be no signal duct at the base of a mountain, a beginning of the duct part way up the mountain, an increasingly good duct near the mountain top (apparently the duct's center), but a complete loss of the duct at the very top. Ducts can almost be visualized as an electromagnetic trough akin to air-conditioning ducts. Operation bears an analogy to two people standing at opposite ends of an air-conditioning duct and conversing through the duct. If either of the people moves either vertically or horizontally out of the duct range, communications would fail. The same effect is found in uhf or microwave ducting. The amateur's challenge thus

Fig. 9-1. This 4-foot parabolic dish antenna, demonstrated by Deborah Franklin, provides quality performance and high gain at 2.3 GHz.

includes locating the duct, trying to determine its path, and getting a signal passed through the duct.

At a lower frequency, the 1269-MHz amateur band exhibits characteristics more closely related to uhf than to microwave. Signals on 1269 MHz, for example, suffer only minor attenuation from small amounts of foliage, etc., and are seldom reflected by less than torrential rains. Reasonable power levels are usually employed on the 1269-MHz band, with 5 to 15 watts being considered normal. The W6ORG Amateur Fast Scan TV repeater in Arcadia, California is one of the higher-powered systems used on this band, and it is known for its outstanding range by amateurs in nearby states. The W6ORG repeater itself is fully solid state, and constructed of many printed-circuit-board modules. These modules, which consist of the receiving preamp, receiving converter, transmitter exciter, rf amplifier, video processor, etc., are available directly from Tom, W6ORG at a very modest cost. In fact, Tom O'Hara received so many requests for his modules, he now produces the units on a large scale (PC Electronics, 2522 Paxson Lane, Arcadia, California 91006).

Still lower in frequency, spectrum characteristics again resemble those of longer wavelength counterparts. Signals on 800 MHz, for example, propagate similar to signals in the uhf spectrum. Communications range, however, is usually greater than that of commercial television counterparts because of greater efficiency and state-of-the-art amateur receiving arrangements. The conventional television receiver seldom uses an rf amplifier stage before its superhetrodyne converter. Whenever rf amplifiers are used, their noise figures are usually high and thus again show poor efficiency. Most amateur receiving front ends, however, are set up to handle low power and weak signals. This, in conjunction with highly efficient antenna designs, is the basis of DX operation at 800 MHz.

Finally, propagation at 70 cm, or 432 MHz, is considered as a point of reference. Propagation on this uhf band is similar to that experience on 2 meters. Only a slight reduction of communications range is experienced, most of that loss being over line of sight paths. The 432-MHz band isn't normally affect by rain, and signals seldom bounce off metal structures in a predictable manner.

SAFETY CONSIDERATIONS

In addition to the obvious and yet most commonly experienced danger of antennas accidently coming in contact with high-tension power lines, several additional factors influence equipment setup within the microwave spectrum. First, always avoid looking directly into the feedhorn or rf-output area of a microwave transmitter—regardless of its rf-power level. Microwave energy exhibits a heating effect (a prime example is the popular microwave oven), and eye tissues are one of the most sensitive areas of the human body. The area of heating effects produced on body tissue by microwave radiation has become a most controversial issue during recent times. Semi-informed individuals and groups, under influence of miniscule, and often misleading, information, have conceived the idea that any and all rf energy presents a critical health hazard. Indeed, several nationwide programs were unsuccessfully launched during 1981 to reduce or eliminate this treacherous offender. The programs, somehow guided by the fact that microwave ovens cook through the process of electromagnetic radiation, and that minimum rf levels in the United States are higher than those permitted in Russia, pushed for some extreme and rather unbelievable measures. For example, it was proposed that any and all transmitters producing electromagnetic radiation (regardless of frequency) should be banned from locations near residential areas. Transmitted signals were to be restricted in such a manner as to prevent exposure of humans, etc. While such measures can truly be considered outlandish, a certain (small) amount of truth can be extracted and should be respected. Electromagnetic energy *can* be dangerous—particularly at high levels and at microwave frequencies. Standing in front of a high-powered radar antenna for a few seconds, for example, may deteriorate bone marrow. Ultimately, this will cause leukemia and probable death. This situation, however, is not the same as sitting in a home 20 or 30 feet beside of beneath an amateur parabolic-dish antenna radiating 500 watts into free space. Likewise, physical harm from electromagnetic radiation in the 3-GHz and lower range, and at power levels under 100 watts, is highly unlikely—unless you deliberately block the antenna for several hours. A condensation, in layman terms, of electromagnetic radiation effect on body tissues is presented in Table 9-1. Studying that table, you may logically surmise that operations within the amateur spectrum present health hazards only when you take extreme measures and/or perform obviously unnecessary activities well beyond the need of normal communications. Properly respected, amateur activities at 14 MHz, 2.3 GHz or 10 GHz are logically justified and environmentally safe. Improperly used, any of these frequency ranges *may* be capable of causing damage. However, the means of such improper use generally require finances which are simply beyond the reach of most amateurs.

EXPANSIONS AND REFINEMENTS FOR MICROWAVE SYSTEMS

Once the radio amateur has successfully established a functioning microwave system and become reasonably familiar with its operation, the next steps will surely involve expanding capabilities and applications. These activities may include anything from adding preamplifiers and larger antennas to completely rebuiding the

Table 9-1. A Chart of rf and Electromagnetic Heating Effects on Human Tissue.
(Information was compiled from details presented in public reports and manuals over a period of years.)

ERP	10 GHz	3 GHz	2 GHz	144 .4 GHz	Below MHz	144 MHz
10 KW	Immediate cooking within 10 feet	Immediate cooking at 2 feet	Immediate cooking at 2 feet	Possible heat at close range	Possible heat at close range	*
2 KW	Cooks within 30 seconds at 10 feet	Cooks within 30 seconds at 2 feet	Cooks within 30 seconds at 2 foot	Cooks within 30 seconds at .5 foot	Possible heat at close range	*
500 W	Cooks within 2 minutes at 15 feet	Cooks within several min's at 1 foot	Cooks within several min's at 1 foot	*	*	**
10 W	May cook over several hours period	Very low heat at 1 foot	*	**	**	***
.1 W	Very low heat over several hrs	*	**	**	**	***
.01 W	Very low heat over many hrs at close range	**	**	***	***	***

*considered negligible except over several-year period at very close range
**no ill effects reported
***no ill effects reported: services used over 6 decades

system to include high-power rf stages, special pulse modulators, etc.

The most logical and cost effective means of increasing communications range is accomplished by the addition of a receive preamplifier. These units are readily available in kit or wired form from a variety of manufacturers. Look for advertisements in monthly amateur magazines for manufacturers and equipment. One such high-performance 2.3-GHz receiving preamp is shown in Fig. 9-2. This particular preamplifier is produced by Universal Communications of Arlington, Texas. It covers the frequency range of 1800 to 2500 MHz. This unit includes a filtered output stage that removes image noise that could degrade the desired signals. The

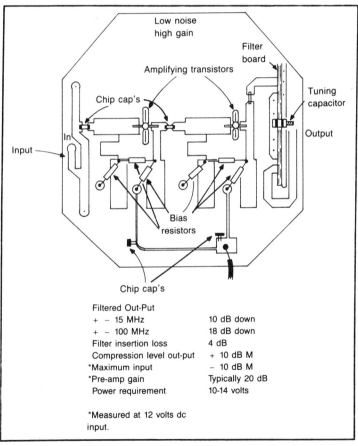

Fig. 9-2. Sketch of high-performance receiving preamplifier that can be mounted with a converter for high gain and low noise.

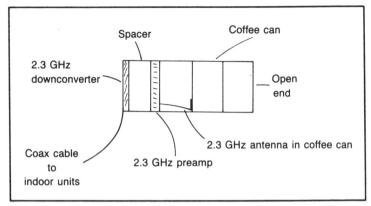

Fig. 9-3. A method of mounting a 2.3-GHz receiving preamplifier and downconverter on the rear area of a one-pound coffee can.

preamp is mounted with its associated 2.3-GHz downconverter on a one-pound coffee can. See Fig. 9-3. Such arrangements are quite popular and efficient for low microwave applications. The preamplifier's gain, incidentally, is typically 20 dB with less than a 2-dB noise figure.

High gain and large microwave antennas are, in a manner of speaking, where one finds them. Parabolic dishes of large area often appear on the military surplus market or occasionally as discarded items around television stations. Considerations when securing such antennas involve primarily its overall area (the larger, the better), and its surface accuracy without significant dents of a wavelength or more at the desired operating frequency. Although relatively expensive if purchased new, satellite-TV receiving dishes on the order of 12 to 16 feet perform extremely well for all amateur operations. Other times, dishes of 3-to 4-foot diameter can be acquired from surplus sales at public utilities (telephone, power, etc.). Numerous amateurs have satisfactorily constructed their own of parabolic dishes from available resources. The structure of these range from wooden struts stretched into a parabola by Nylon ropes, to sheets of scrap aluminum folded into a corner-reflector form. The key to success in this area is keeping a sharp eye out for applicable items.

Chapter 10

Interfacing Microwaves With Television and Computers

The combination of amateur television and microwave setups is an almost natural marriage, as both systems exhibit broadband concepts and adaptivity to expansion. Because minor interaction between the popularly used amateur microwave bands are usually experienced, microwave repeaters or links that relay video signals have also proven highly feasible. The combination of these two aspects thus indicates a very promising frontier for the amateur. One of the most commonly raised questions in this area may explore the benefits of using microwave rather than uhf frequencies for fast scan television operations. Because lower-frequency ranges, such as the established 70-cm band, provide good distance and occasional ducting capabilities, it might seem the use of higher frequencies wouldn't be advantageous. Such isn't necessarily the case; the line of sight (microwave) frequencies can be linked for long distance communications, or relayed via satellite for almost worldwide coverage. Additionally, the microwave frequencies afford highly dependable and predictable communications abilities. Lacking conventional propagation variables associated with lower frequencies, the microwave spectrum is destined to prove its merit as a reliable television carrier.

The usual video-modulation techniques used at 1.2 and 2.3 GHz are conventional amplitude modulation in nature; however, frequency modulation is typically used above 4 GHz. Reception of commercial TV signals in the 4-GHz range is accomplished with dedicated receivers that amplify, detect, and convert the video signal to baseband (30 Hz to 4 MHz) before applying it to an rf modulator (small transmitter) for application to the television receiver's antenna terminals. Amateur video reception at 10 GHz usually depend on slop-detection arrangements, proven in amateur activities for many years. As technology advances, we may see SSB and numerous digital-based systems replace that old standby.

FAST-SCAN TV AT 2.3 GHz

The large number of manufacturers producing television downconverters for receiving MDS transmissions in the 2100-MHz band almost has this range ready and waiting for the amateur microwave enthusiast. These relatively inexpensive converters mount to the bottom of a one-pound coffee can that is mated with an associated high-gain antenna. The usual frequency range of these converters is 2000 to 2500 MHz, the precise frequency range being selected by trimming of the local oscillator circuit's strip line. The usual i-f output bandwidth of these converters is approximately 30 MHz, reaching from commercial TV channel 3 or 4 through channel 6. This is essentially a broadband system, although a specific center frequency is determined by oscillator strip length. The overall system provides the capability of tuning approximately 30 MHz at 2300 MHz by selecting various i-f ranges. The converter proper is a linear device; it does not change any form of modulation arriving at its input. As a result, FM audio and A-M video signals can be processed directly by the unit. The amateur contemplating ATV operations at 2.3 GHz needs only connect the converter's output to a standard television receiver for quick and easy fast-scan TV reception. Accompanying amateur TV transmitters for 2.3 GHz are becoming quite popular and readily available from many converter manufacturers. Amateurs who want a first-class setup can include a vestigal-sideband filter on the transmitter output for broadcast-quality transmissions. In order to enjoy the full effect of this 2.3-GHz ATV activity, an audio subcarrier mixer/processor is highly desirable.

FAST-SCAN TV VIA 10 GHz

The concepts associated with amateur video communications at 10GHz may be placed in one of two categories: simple or sophisticated. The simple approach is sufficient for most amateur applications, and it offers the advantage of low cost and relatively quick assembly. Conversely, the sophisticated approach resembles concepts utilized in satellite TV systems (TVRO) and is quite professional as compared to usual amateur applications.

It is quite apparent that the bandwidth required for television-signal transmission reduces signal strength, and consequently decreases effective communications range. As a relatively quick calculation, consider the signal to decrease approximately half (3 dB) with each doubling of bandwidth (using a voice-system 200-kHz bandwidth as reference). A 1-MHz bandwidth signal to 10 GHz may thus be considered 15 dB below a 200-kHz audio link over the same path. These parameters should not cause one to abandon the television communications idea, however, but simply realize that reduced distances are inherent with increased bandwidth. As a rough estimate, an approximate 20-mile audio path will probably reduce to 5 miles for video signals.

Simple System

The simple technique for amateur television communications at 10 GHz involves using Gunnplexers in a modified TV-oscillator type arrangement capable of extended range. This technique is shown in Fig. 10-1. Baseband video, along with (optional) audio, is applied to the Gunnplexer's varactor diode. The other unit receives these broadband signals and outputs them (at its mixer-diode terminal) to an unmodified television set. As in the case with 10-GHz audio links, a dc bias is superimposed on the modulating video signal to establish a precise operating frequency. This frequency will be determined by the i-f frequency (channel to which the TV receiver is tuned). With a 54- to 60-MHz i-f, for example, one Gunnplexer's varactor-diode bias would be set for operation on 10.054 MHz. The resultant frequency offset (540-60 MHz) provides a channel-2 signal for the television set. If operation on another channel is desired, varactor-diode bias may be adjusted to provide that frequency. It's that simple! While on the subject of various i-fs, bandwidths, etc., it should also be noted that more than one video signal may be relayed simultaneously by a 10-GHz Gunnplexer system. Two or three channels of information may be multiplexed on the same 10-GHz carrier by using a different varactor diode bias to specify each signal's transmit and receive frequency, as shown in Fig. 10-2. Seven-volts bias is used for video-signal 1 to provide a channel-2 received signal; 12-volts bias is used for the second video signal (channel 3); and 17-volts bias

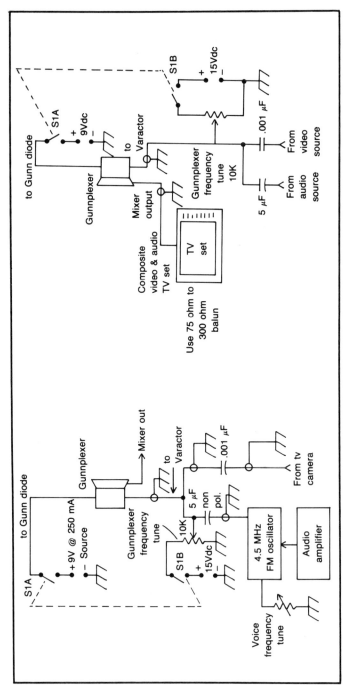

Fig. 10-1. Basic arrangement for using 10-GHz Gunnplexers for short-range TV communications. The two frequencies are offset by amount corresponding to the channel viewed on the TV receiver.

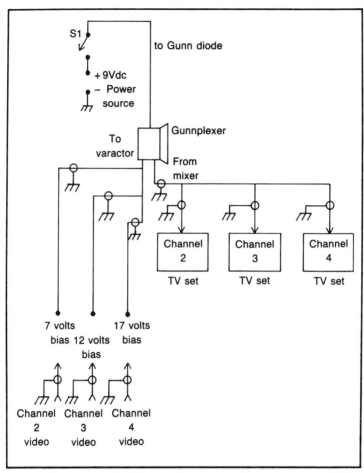

Fig. 10-2. Two or three television signals can be multiplexed on a single 10-GHz signal. Different bias voltages are applied to the varactor diode for each TV channel signal. This setup represents one side of a system; the other side is identical in layout.

is used for the third video signal (channel 4). Because signal strength decreases inversely with bandwidths. The inclusion of additional TV channels can be logically expected to significantly reduce system range. As an example, we can expect the previously discussed 5-mile-range video system to decrease to approximately one mile when several multiplexed channels are employed. However, the advantages of such links are many, and usually worth the distance losses.

Techniques for 10-GHz video operations are similar to audio

operations, with only a couple of exceptions. The transmitting unit varies varactor-diode bias until the received signal is present on the desired TV channel. The television set's fine tuning is then used to follow long-term drifts. Few additional problems should be encountered, provided both operators understand that mutual and simultaneous tuning (like trying to make two motors run at same speed) is not possible.

SCAN-CONVERTING RELAYS

An interesting concept that holds unparalleled appeal in amateur applications involves using microwave links in television scan conversions. Using this arrangement, several amateurs in an area can realize very wide range from their ATV setup. A block diagram is shown amplified in Fig. 10-3. A 1269-MHz amateur television receiving converter is used in front of a conventional television set for acquiring amateur video signals, and a 2.3-GHz transmitter is used for relaying amateur video. A slow-to-fast-scan digital scan converter (such as the popular Robot 400, etc.) is used in conjunction with an h-f band transceiver, completing the low-band portion of this setup. The combination of carrier-operated relay, tone decoders, and audio-subcarrier generator combine the individual units into a full-blown system. In operation, the system can be used as a fast-scan TV repeater, or as a slow-to-fast and fast-to-slow scan converting system that provides worldwide communications via a high-frequency band such as 20 or 10 meters. The popular frequencies for slow-scan TV communications, 14,230 kHz and 28,680 kHz, are suggested for fixed-frequency operation. The advantage of remote tuning, however, should prove quite beneficial. A few minutes study of the illustration should reveal many unique capabilities. Fast-scan TV signals arriving via a relatively inexpensive 1269-MHz amateur converter are presented to the television receiver for video and audio demodulation. The video signal is directed to the digital scan converter and simultaneously relayed on 2.3 GHz for monitoring via inexpensive receive converters. A voice-operated relay is used to disconnect slo-scan converted signals from the h-f transceiver during audio transmissions on 1269 MHz. Reception of long distance video is accomplished through the h-f transceiver (activated via a tone switch). These signals are loaded into the digital scan converter, and retransmitted via 2.3 GHz to the fast-scan operator. This operator has both audio and video capabilities, and has the option of listening to SSTV signals during reception. The beauty of this particular arrangement lies in portability of the operating stations. A basic setup consists of a TV camera, 1269-MHz transmitter, 2.3-GHz downconverter, and an unmodified television receiver.

Numerous expansions fo this setup are, of course, possible. The general setup presented is merely a guideline for amateur expansion.

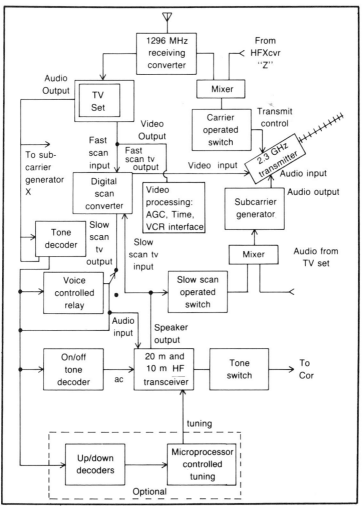

Fig. 10-3. Block diagram of a slow-to-fast and fast-to-slow scan-converting relay/repeater system. Video processing unit provides on-screen readout of signal strength, time, and directs VCR tapes to scan-converter input.

LINKING HOME COMPUTERS VIA MICROWAVES

Home computers can be interlinked by microwaves in a relatively inexpensive manner. Such links may be considered from one or two standpoints: directing computer video to a microwave link for high-performance modulated-oscillator type transmissions, or direct-communications interfacing such as through RS232-compatible modems. Computer-video linking follows the same concepts outlined in the 10-GHz video discussion. Essentially, this arrangement involves directing computer-display video to an amateur transmitter and tuning that signal on an accompanying receiving downconverter. Although this is a form of slop detection, results are gratifying and usually sufficient for amateur applications.

This chapter's technique of computer communications linking via microwaves involves directing data lines to a modulator or modem, and then to the transmitter. The appealing aspects of this technique are flexibility and adaptivity to expansions. The basic computer communications setup is shown in Fig. 10-4. Simply, stated the microwave link replaces conventional telephone lines. Although microwave frequencies are now not mandatory for such transmissions, this will not be the case within a few year's time. As faster data speeds become popular, higher baud rates will become necessary. As communications progress beyond 1200 and

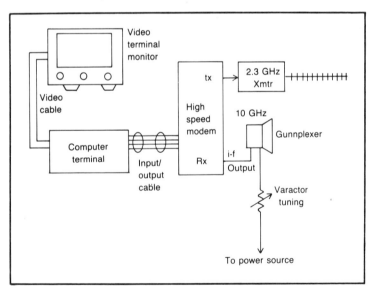

Fig. 10-4. A basic duplexed microwave system for computer linking. The system can host a variety of modem speeds.

2400 baud to 4800, 9600, and higher, bandwidth expands to the point of requiring use of microwave frequencies. A farsighted plan in that direction is thus quite beneficial for amateur installations.

Another arrangement involves placing a semi-intelligent printer on a microwave link accessible to its mutual users. Through the use of memory buffer and a compatible modem, each amateur can access the printer as desired. Additional concepts are unlimited; I leave their development to you amateur experimenters.

Chapter 11

Amateur RADAR and Intruder Alarms

Can an amateur living in a remote or isolated area enjoy involvement with microwaves? Indeed he can, and the possibilities are many. Some examples and design outlines are discussed in this chapter. Amateur RADAR systems are capable of maintaining a watchful eye in some rather interesting ways, ranging from nightwatching to following storm cloud activity. Their range may be somewhat restricted compared to commercial equipment, but these personal applications can fill a void left by such professional and more expensive units.

Although RADAR systems have been in existence many years, their use in individual or amateur setups has been miniscule. This situation could be because their applications haven't yet been brought to light. Today's escalated concern for both mass and individual safety, however, has radically altered that situation. People are concerned with happenings around them to the extent of using personal RADAR. Some of these units take the form of motion sensors used in open areas, etc., while others merely pass a RADAR beam in a set direction to sound an alert if intruders enter the watched area. Additional personal RADAR systems are also being pioneered, but very little information concerning their operation has been released at the time of this writing.

RADAR is an acronym for RAdio Detection And Ranging: a technique that uses a beamed radio wave (usually in the microwave spectrum) that reflects off metallic objects. Through precise timing of transmitted pulses and echoes, exact distances can be calculated. A simple form of RADAR is used by bats: their shrill cries reflecting off walls and cliffs provide these animals with distance references in a crude, but highly effective manner.

RADAR TYPES

The most popular form of RADAR is the type used by military operations. These systems use precisely timed pulses which can be used in a highly predictable manner. The RADAR beam, in this respect, acts as a form of invisible light beam, surveying designated sectors of an area. The results are displayed on a long-persistence cathode-ray tube which is usually calibrated in both direction (azimuth) and distance (range). Radio waves travel at the same speed as light waves: 186,000 miles per second, or 300,000,000 meters per second. Translated into more usable terms, a radio wave travels one mile in 5.3 microseconds. Microwave frequencies are used for RADAR because of their line-of-sight propagation and precise directivity. Thus, by transmitting a narrow-beamwidth microwave signal in a specific direction, and precisely timing the received echo, the distance to the object in that direction can be determined.

One of the most common forms of RADAR is the PPI, or plan position indicator. This system, which displays an electronic map of the surrounding area, utilizes a rotating parabolic dish antenna synchronized to a rotating yoke on the cathode ray tube (crt). Each sweep of the crt electron beam begins at the center of the screen when a transmitted pulse emanates from the antenna. The transmitter then shuts off to permit the receiver to listen for echoes. Meanwhile, the crt electron beam continues scanning outward from center, and any echos are displayed as intensified areas on the screen. The process continues with rotation of the antenna and crt yoke until full 360-degree coverage is achieved. The complete operation is then repeated on the next sweep.

Another widely known form of RADAR is doppler shift, or police-gun RADAR. These units operate on the principle that microwave energy reflected by a moving body is shifted in frequency. The amount of this shift is directly proportional to the object's speed compared to the transmitting unit (the faster the speed, the greater the frequency shift). An example of this is a train whistle moving past an observer. The whistle's frequency is high as the train approaches the observer, and shifts lower as it moves away from the observer.

Police doppler RADAR functions similar to a Gunnplexer. A

continuous signal is transmitted on 10.5 GHz; the reflected signal is also received by that unit, and produces a difference frequency that is directly related to speed of the target. Low speeds produce small frequency shifts, and high speeds produce greater frequency shifts. Police doppler RADAR usually have what is called a zero i-f, because audio-range frequencies are employed for speed measurements. See Fig. 11-1. The general outline of a doppler RADAR is shown in Fig. 11-2. An audio-frequency signal is applied to a Gunn oscillator and transmitted toward a moving object. The reflected and doppler shifted signal is received by the same Gunnplexer unit and directed from its mixer output to a high-gain amplifier. The original signal and the reflected signal are then compared, and their difference is converted from analog to digital for direct readout. The display section includes interpolation of frequency to speed, providing miles-per-hour indications. A similar amateur doppler RADAR can be devised by simply applying an approximately 1,000-Hz modulating signal to a 10-GHz Gunnplexer and monitoring the returned signal via an audio amplifier connected to the mixer output terminal.

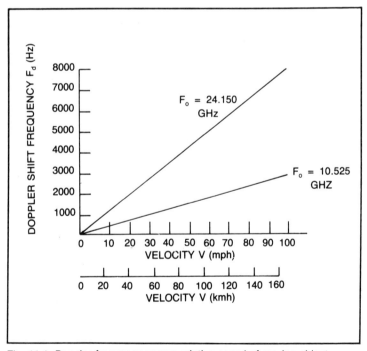

Fig. 11-1. Doppler frequency versus relative speed of moving object.

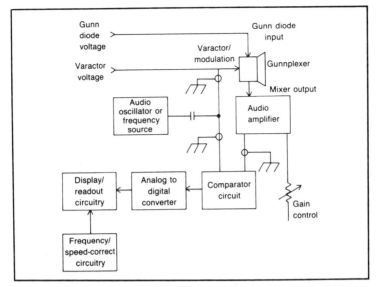

Fig. 11-2. Block diagram of a doppler RADAR system similar to the police RADAR guns. Units of this type can be constructed using the popular Gunnplexer. The audio frequency source may be a simple NE555 circuit, or equivalent.

The recently uncovered phenomena of police RADAR guns "clocking" trees or buildings at 60 or more mph has been traced to a rather interesting culprit. Wind blowing through damp leaves on a tree can cause rapid undulations, producing extreme doppler shifts on reflected signals. Buildings containing large machinery often experience minor but rapid vibrations of walls and flooring. When the RADAR gun is pointed toward that building, the wall fluctuations produce high-speed indications.

INTRUDER/MOTION-DETECTOR ALARMS

Ten-GHz transceivers, such as the popular Gunnplexer units, can be adapted for motion-detection and alarm use in a very effective manner. This technique is similar to that of doppler-shift RADAR, except dual mixers and output amplifiers are used in conjunction with a phase detector, similar to the setup shown in Fig. 11-3. The output of the phase detector connects to a differential amplifier that is adjusted to ignore normal background movements (such as fan blades moving, etc.). Random movements will then produce an output from the differential amplifier, which can be used for activating an alarm circuit.

Microwave intruder alarms have been used for several years. Unlike more conventional infrared systems, microwaves can be scattered over a relatively large area. The sensitivity of these units can be set to detect non-metallic objects such as furniture, etc. Suitable dual-mixer microwave units are available from Microwave Associates of Burlington, Massachusetts.

10-GHz MINI-RADAR CONCEPTS

There are times when a radio amateur has a need for a basic RADAR system to monitor activities in or near his house, or for surveying traffic flow on nearby roads. At other times, a beach cottage or mountain cabin can be afforded added security whether occupied or not. Beach areas are ideally suited for amateur mini-RADAR; their flat, unobstructed nature affords maximum range and sensitivity for such setups. A system of this nature could also appeal to amateur boating enthusiasts.

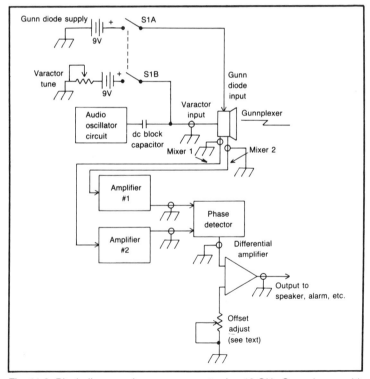

Fig. 11-3. Block diagram of an arrangement using 10-GHz Gunnplexers with dual mixers for intruder/motion-alarm circuit. The "offset" control compensates for normal variations within area being watched.

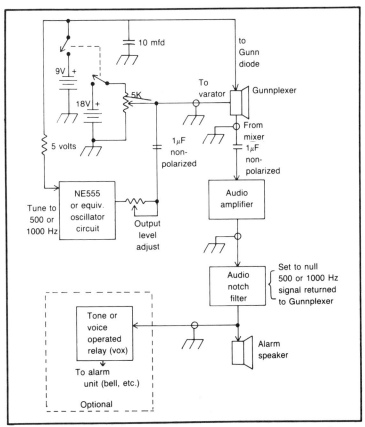

Fig. 11-4. Block diagram of a mini-RADAR system for personal use. System works by nulling normally reflected signals. Unexpected movements offset that null, causing audible tone. A suitable notch filter for this circuit is the MFJ751, sold by MFJ Enterprises, Box 494, Mississippi State, Mississippi 39762.

A mini-RADAR may be constructed in several ways, depending on desired applications and available funds. A beginning setup is shown in Fig. 11-4. This system is based on doppler-shift principles for simplicity. Battery operation is provided for portable use if required. An audio signal generated by an NE555 integrated circuit is used to modulate a 10-GHz Gunnplexer. This signal is transmitted in a directional or omni-directional path, according to desired coverage, and returned to the Gunnplexer's mixer. The output of the mixer stage is directed to an audio amplifier (such as the audio section of a portable A-M radio; merely connect the mixer output cable to the volume control wiper and ground, and disconnect prior stages). The audio amplifier output is then connected to a tunable notch filter capable of providing at least 50 dB of attenuation. The notch filter should be tuned to the same frequency as the modulating tone (perhaps 500 or 1000 Hz). An alarm may be connected to the notch filter's output, if desired.

During operation, the Gunnplexer illuminates the selected area with microwave energy which is returned to the mixer in a normal manner. This audio signal is then nulled by the notch filter, producing zero audible output. Whenever unexpected movements occur in the watched area, or foreign objects appear (increased signal amplitudes), the audio tone can be heard. The audible tone is usually sufficient for warning of intrusions; however, a common voice-operated relay can be included to switch on lights, recorders, etc., if desired. Physical mounting arrangements for the Gunnplexer will be dictated by system capability (such as pointing upward into a dish for omni-directional use, pointing horizontally for area monitoring, etc.). I leave that situation open to your ingenuity. Expansions of this basic setup are possible; I've only outlined basic concepts. Now set your creativity to work and design the ultimate amateur mini-RADAR.

10 GHz AMATEUR WEATHER RADAR

The applications of commercial weather RADAR systems have truly proved their worth during recent years. In addition to applications for local and national weather forecasting, weather RADAR systems have also acquired widespread popularity during periods of bad weather. The number of lives saved by commercial television broadcasts of weather RADAR is unknown, but everyone is surely familiar with the many benefits of these systems. One possible shortcoming of commercial weather RADAR systems might be their lack of availability at specific times, or inability to scan a particular small area (such as your own neighborhood). Bearing this in mind, the author set about designing his own type of personalized weather RADAR: an inexpensive and easy-to-construct setup with a relatively short range for scanning local activity. The project has been progressing in fine style; however, a number of additional and more important projects have curbed final refinement of this system (each day is, unfortunately, restricted by a 24-hour time limit). Rather than leaving the almost complete project dormant, I will pass along my notes thus far as guidelines for others interested in pursuing or refining this concept. Please bear in mind, however, this is not a complete built-by-the-numbers project.

This system is based on the principle that a radio signal travels one mile in 5.1 microseconds, and that an electron beam can be caused to travel across the screen of a cathode ray tube at a predetermined rate. If a radio wave leaves a transmitting antenna at the same time a horizontal line begins on the left side of a crt display, for example, the wave will travel approximately 12.1 miles in approximately 63 microseconds (the typical time for scanning a horizontal line in a conventional television set).

Version One

The basic unit is shown in Fig. 11-5. An oscilloscope with a horizontal trigger output is used in conjunction with a 10-GHz Gunn-

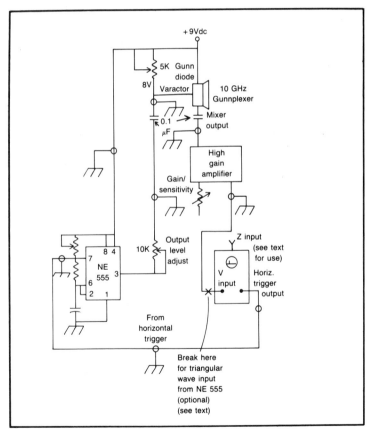

Fig. 11-5. Block diagram of the first generation amateur weather RADAR system by K4TWJ. The Gunnplexer transmits and receives 10-GHz energy, and the oscilloscope provides timing and range information.

135

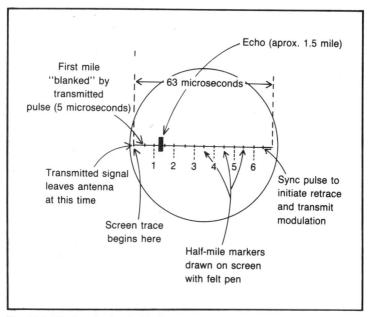

Fig. 11-6. Display obtained with the weather RADAR shown in Fig. 11-5. The scan begins at the left of the screen, at the same time a modulating pulse is applied to the transmitter. The time periods shown will provide an approximately 6-mile range.

plexer, an NE555 pulse-modulating circuit, and a high-gain amplifier to produce a direction-and-distance type weather RADAR. Although the Gunnplexer is usually employed in doppler-type systems rather than pulse-timed systems, the arrangement shown works quite well. The Gunnplexer's mixer used the constant transmitted signal as a local oscillator, thus eliminating requirements for T/R (transmit/receive) devices. A horizontal-scan-initiating sync pulse from the oscilloscope's horizontal-trigger output keys the NE555 oscillator, which in turn supplies a modulating signal to the Gunnplexer's varactor-diode input. The Gunnplexer transmits this signal toward the distant object. If a reflection occurs, the return signal reaches the diode mixer, is heterodyned, and fed to the high-gain amplifier. Following amplification, the signal is directed to the oscilloscope's vertical input for display. Since the scan started during the pulse-transmission time, the signal's round trip delay time is indicated by echo placement on the screen. A clarification of this operation is shown in Fig. 11-6. The Gunnplexer modulation should be between 200 kHz and 1 MHz: that frequency determines pulse width and echo-signal width. Using the formula $F = 1/T$, a 200-kHz

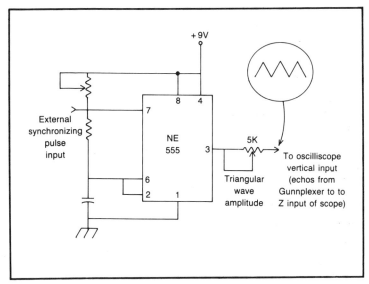

Fig. 11-7. A triangular-wave generator for vertical deflection in sync with Gunnplexer movement.

modulating frequency translates to approximately 5 microseconds, which is approximately a mile on the screen's display. Comparatively, a 1-MHz modulating signal encompasses approximately 1 microsecond. Although the Gunnplexer continuously transmits a carrier, only the brief modulating pulses cause indications on the monitor screen. A 200-kHz modulating signal (5.1 microseconds) will block the first mile of screen display; however, this hasn't been a problem.

This basic system provides distance indications and rain-cloud density information (according to the amplitude of the echoes), but does not include directional information. Since this system's prime purpose is watching an area not covered by commercial weather RADAR, or tracking local storms (which, in most cases, usually arrive from a known direction) the lack of directional information is a justified trade-off. The second version of this mini RADAR, however, includes directional information.

Version Two

The second version uses a modified (fast speed) oscillating-fan motor mount to provide Gunnplexer movement over an approximate 170-degree area. An NE555 triangular-wave generator provides vertical deflection of the crt beam to coincide with Gunnplexer movement. A circuit for this function is shown in Fig. 11-7. Synchronizing pulses can be applied to pin 7 of the NE555 to initiate retrace and scan. This pulse can be acquired from a microswitch mounted at one end of the Gunnplexer's mounting platform. The triangular wave circuit is connected to the oscilloscope's vertical input, and the echo amplifier output is then connected to the Z, or intensity, input. See Fig. 11-5. Careful readjustment of amplifier gain and oscilloscope intensity should produce a dark screen until an echo is displayed. Range markers can be electronically added to the display by another NE555 circuit, or they can simply be drawn on the screen with a felt-tip pen. As mentioned, various oscilloscope sweep rates may be employed for monitoring distances according to your needs. Because the Gunnplexer is looking upward, its maximum range should be obtainable. I suggest, however, holding that distance to less than 15 miles. It's also advisable to replace the P1 phosphor crt with a long-persistence P7 phosphor equivalent.

During operation, synchronizing pulses from the oscilloscope trigger the NE555 modulation circuit, causing the Gunnplexer to transmit a 200-kHz signal. As the scanning line moves across the oscilloscope screen, the 10-GHz signal travels out and reflects off clouds (according to their density). The returned echos beat against

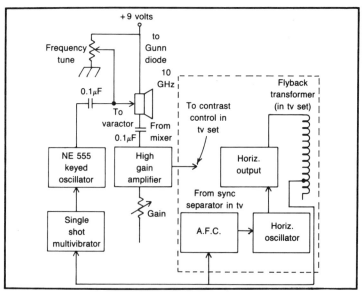

Fig. 11-8. The "third version" amateur weather RADAR uses a modified television set for timing and display of target information.

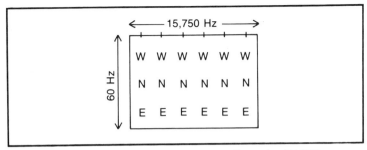

Fig. 11-9. Display of the complete "third version" weather RADAR as seen on the screen of a modified TV set.

the Gunnplexer's carrier, are boosted by the high gain amplifier and applied to either the oscilloscope's vertical input or Z-modulation input (depending on the design). Other refinements can be added to this setup; I'll leave those expansions to your creativity for developing a truly personalized system.

Version Three

A third version mini RADAR is also possible. This system could employ a television set rather than an oscilloscope. TV receivers are often available quite inexpensively from repair shops. Units with defective tuners and/or i-f stages are especially appealing. Because the RADAR signal is injected at the video amplifier, prior stages are not required. An outline of version three is shown in Fig. 11-8. Horizontal afc pulses from a low-voltage winding on the TV flyback transformer can be used to trigger the Gunnplexer's modulator (these pulses are extremely accurate and quite narrow in width). They maintain horizontal sync in the television, and provide RADAR timing. A Schmitt trigger or single-shot multivibrator circuit should work, or the flyback pulses can be applied directly to the NE555 (this agc keying level should not exceed 6 volts. A dropping resistor may be required with some TV sets).

Output from the modulator circuit should be applied to the Gunnplexer's varactor diode. Reflected 10-GHz signals are boosted to 2-volt level and applied to the TV video amplifier (the contrast control wiper and ground are a logical point for this signal insertion). Amplified echoes then appear as intensity variations on the screen, with timing and distance references similar to version two. The next step in system development will include electronic panning of the 10-GHz signal to match the TV set's vertical scan rate. An outline of that display is shown in Fig. 11-9.

Version Four

A fourth generation mini weather RADAR is presently on the drawing board, and it promises to be a truly professional unit. Essentially, this system will use an inexpensive microcomputer such as the popular TRS-80 Color computer, Commodore VIC-20, or such, for generating timing signals and for providing a color display. Microcomputers are quite useful for such applications because they are easily programmed for various timing and ranges, for color displays according to signal intensities, etc. The recent inclusion of analog-to-digital and digital-to-analog converters in these units provides readily available interfacing to RADAR electronics.

This project can obviously continue to grow. I trust sufficient information, operating parameters, and ideas have been presented herein to inspire some pioneering in this unique area. I'm certain you'll enjoy pursuing this RADAR concept, and I look forward to hearing of your endeavors.

Chapter 12

Microwave Exclusive: TVRO and MDS

Any book detailing the exciting frontier of microwave communications would not be complete without a discussion of the popular commercial-satellite bands and activities within that spectrum. Two of the more popular areas of interest are the satellite broadcast band of 3.7—4.2 GHz which is used for relaying an extensive amount of video programming to cable networks and private systems, and the MDS or low-power TV broadcast band of 2.15 - 2.18 GHz. Thanks to the recent flurry of relatively inexpensive home-satellite-TV receive-only systems (TVROs), this activity has enjoyed nationwide enthusiasm and excitement. This chapter's discussions will, consequently, begin with the popular satellite-TV area.

The ability to acquire satellite-relayed television programs through the use of relatively inexpensive receiving terminals marks the dawning of a new era in video activities. This era shows promise of establishing new capabilities and opportunities for both radio amateurs and small businessmen in a truly exciting manner. The future of conventional television broadcasting will, indeed, be directly related to the satellite field and its operations. Cable TV companies will continue their use of satellites while also embracing Talk-back features (known as interactive cable) allowing viewers to participate in opinion polls, and the like. Another feature, known as videodate or viewdata, will link subscribers with a master computer for home shopping, purchasing, banking, etc., while sitting in the comfort of one' home. Through the use of TV satellites, these functions will expand nationwide, and perhaps worldwide.

Independent satellite-TV receive-only terminals (TVROs), already gaining extreme popularity, will become yet more commonplace for individual and small-business use. These backyard receiving terminals are presently providing their owners with up to 100 channels of specialized TV viewing, while also acting as the heartbeat of low-power microwave TV systems (MDS) to be described later in this chapter. Additional satellites will complement the numerous units presently in operation around the globe, and new satellites specifically designed for direct broadcast to homes will become popular and escalate national telecasting from a vantage point in the skies. Innovations and expansions are unlimited, with applications that can truly stagger the imagination.

THE TELEVISION BROADCASTING SATELLITES

All of the popularly used TV satellites are situated in a parked, or geostationary, orbit located approximately 22,000 miles above the earth's equator. This group of satellites comprise what is known as the Clarke Belt, so named in honor of the science-fiction author and visionary of the 1940s. Because the satellites directly follow earth's rotation, they appear absolutely stationary in the sky (movement, or roll, is less than 1 percent!). Satellite frequency bands are 5.9 to 6.4 GHz uplink and 3.7 to 4.2 GHz downlink. Most of the satellites employ 24 transponders, or separate channel relays, and each channel is 40 MHz wide. A quick check in mathematical calculations indicates only 12 channels would fit into a 500-MHz spectrum; however, a combination of horizontal and vertical signal polarization is used to overcome this obstacle. Although channels of similar polarization are adjacent, cross modulation is not created. This is because bandwidths are typically 30 MHz, and an extra 10 MHz is used for guardband. The video format is wideband FM, with an overall frequency deviation of 30 MHz, which provides a very good signal to noise ratio. TV-satellite audio is conveyed as a subcarrier on either 6.2 or 6.8 MHz, depending on the particular transponder being received (6.8 MHz is presently used on 75 percent of the transponders). Because the maximum clusters of signal energy are produced near the carrier resting frequency, transponders are offset 20 KHZ between vertical and horizontal polarizations. Thus, earth-based setups receiving a particular satellite would shift both polarization and frequency when shifting transponders. The prime workhorse of TV satellites is Satcom 1 which is placed approximately 135 degrees west (in reference to the zero meridian), and carries almost full transponder loads 24 hours a day.

HOME SATELLITE-TV RECEPTION

Extensive use of commercial satellites for relaying vast amounts of TV programming to earth-based ports and cable-TV corporations has led the way for today's popular satellite receive-only terminal (TVRO). Reflecting designs of their substantially more elaborate commercial counterparts, these units consist of a parabolic dish, low noise amplifier (LNA), and satellite receiver. The output can be baseband video or TV-ready signals. See Figs. 12-1, 12-2, and 12-3. Original prices of home TVROs were in the $10,000 to $15,000 range, but technical innovations and other outside influences recently dropped those prices to near $3,000. Currently, additional developments have brought prices to near $2,000.

Fig. 12-1. Twelve foot parabolic dish antenna with 3.7-4.2 GHz downconverter located at the feed (focal) point. This KLM system provides high-quality, noise-free pictures from numerous satellites. A polar mount with motorized tracking affords easy dish movement to acquire various satellite signals (courtesy KLM Electronics, Morgan Hill, California).

Fig. 12-2. Rear view of the parabolic-dish mount shown in Fig. 12-1, showing the moto-trak system for automatic satellite selection. Additional gears in the lower box are not visible (courtesy KLM Electronics, Morgan Hill, California).

The legality of home TVRO operation is presently somewhat vague. Satellite operators and users sensibly do not want intruders eavesdropping on their revenue-producing operations, and the FCC hasn't declared a hard-and-fast rule for all situations. The individual using his TVRO strictly for personal enjoyment argues his point: "The signal is on my property, I can justifiably view it—otherwise, they can remove the signal." Officials occasionally intervene and reflect other opinions. The acceptable use, as of this writing, is to keep home TVROs strictly for personal use, and do not relay the programs to neighbors.

A typical home-satellite-TV receiving setup is illustrated in Fig. 12-4. The large parabolic dish antenna provides gain, approximately 40 dB at 4 GHz. In order to achieve this gain, parabolic accuracy must be maintained to with 1/8 inch tolerances. In order to receive different satellites, the dish must be movable in azimuth and elevation. Reception of various transponders aboard a specific satellite requires rotation of the LNA feedhorn that at the dish's focal point. The LNA is required to provide an additional 30 to 50 dB of gain with a 1- to 2-dB noise figure. There are no logical substitutes for extensive gain and noise figures: satellite signals are cloaked from mass public reception by their microwave frequencies and their low-level signals. Receiving terminals must be capable of digging these signals out of the noise by a minimum of 10 dB. The dish and LNA combination provides this ability. The LNA utilizes GaAsFET transistors to achieve low noise figures. These devices are quite expensive and fragile. Additionally, the GaAsFET LNA must be mounted at the dish's focal point—a location that subjects it to extremes of weather, lightning spikes, etc. The cable connecting the LNA to the indoor receiver is low-loss coax. The indoor receiver is a single- or dual-conversion unit, usually exhibiting 50 to 60 dB gain with a 10- to 15-dB noise factor. Bandwidth is typically to 20

Fig. 12-3. Interior view of the moto-trak system, showing some of the gear assembly (courtesy KLM Electronics, Morgan Hill, California).

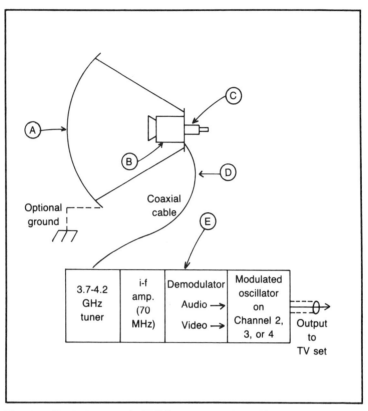

Fig. 12-4. Block diagram of a TVRO setup, showing at (A) the parabolic dish antenna; at (B) the LNA; at (C) the rotor to change polarization of rf-signal pickup element; at (D) the coaxial cable to feed dc operating voltage to the LNA and to route the microwave signals to the tuner; and at (E) the indoor tuning unit.

to 30 MHz, and audio detectors operate on 6.8 or 6.2 MHz. The output of this (24 transponder, 500-MHz range) satellite receiver is baseband audio and video. Signals from the satellite-receiving setup are then processed and connected to the TV set. See Figs. 12-5 and 12-6.

MDS: WHAT IT IS

Another of the more recent forms of restricted television broadcasting involves the rapidly growing medium of MDS; a microwave system for carrying pay-TV type programming to subscribers within a particular metropolitan area. This service may be considered a cross between conventional television broadcasting and cable-TV operations with some notable exceptions. Only one channel of television programming is used, and that programming usually involves subjects not included in normal broadcast TV coverage. Because of the ability to serve this void, MDS systems are springing up around the country in great numbers.

MDS is an acronym for Multipoint Distribution Systems: a low-cost, common carrier designated for point-to-point communications of various independent services. The system, which is allocated in the 2150- to 2180-MHz (2.15- to 2.18-GHz) microwave range, was originally established by the Federal Communications Commission as a form of commercial transmitting service for relatively narrow-band signals. When the demand for this service fell short of its predicted goal, the FCC reapportioned MDS transmitter/channel bandwidths to 6 MHz for inclusion of NTSC television signals. Thus began the saga of low-power microwave broadcasting of pay-TV programs. The microwave concepts of pay TV took several years during the latter 1970s to become popular. From a beginning as MDS relays of special satellite programming and video tape recorder runs of restricted movies to hotels and motels, the concept grew to include numerous metropolitan broadcasts of similar programs for subscribing home viewers. Additional MDS broadcasting services, many instigated by individuals rather than larger corporations, joined the action and began broadcasting to specific audiences in each metropolitan area. Activity soon flourished as the microwave-TV broadcasting craze gained an almost immediate foothold in densely populated areas. Suddenly, the vision of one establishing his own revenue-producing TV station was strikingly

Fig. 12-5. The KLM Sky Eye IV satellite TV receiver, with remote downconverter on top of the unit. This is a state-of-the-art unit priced in the range of many familiar Amateur Radio high-frequency transceivers (courtesy KLM Electronics, Morgan Hill, California).

Fig. 12-6. The KLM 11-foot parabolic dish and motorized mount provides excellent performance for almost any area of the United States (courtesy KLM Electronics, Morgan Hill, California).

close to reality. The bare-bones setup for many of these neighborhood TV stations consisted of a satellite-TV receive-only terminal (TVRO), a video tape recorder/player and possibly a basic TV camera for brief live announcements. The MDS transmitter proper was remotely located and maintained by a licensed owner. An early FCC stipulation required MDS transmitter ownership to be different from its user; a leased-transmitter arrangement. As this book is being written, MDS arrangements continue their phenomenal rise in popularity with little indication of slowing down.

OPERATIONAL CONCEPTS OF MDS

The arrangement for MDS transmissions of pay-TV signals consists of a satellite-TV terminal, and sometimes a studio containing one or more video-tape players which feed video and audio information to the MDS transmitter. Either an omnidirectional or directional microwave antenna is employed, depending on the specific area to be covered and the MDS transmitter location in relation to that area. Microwave signals (specifically, 2.1 GHz for MDS operations) are strictly line of sight, thus restricting transmitter distance to a normal 20- to 25-mile range. Exact distance depends on terrain and transmitter height, however, diminishing field intensity usually requires expensive (at least for MDS operators) receiving setups at the outer limits of coverage. This is a cost-effective game: Receive terminals must necessarily be dependable, but they must also be inexpensive if a reasonable income is to be realized.

While the concept being outlined is cloaked from mass public reception only by its difficult-to-receive microwave range and its coverage restrictions, additional scrambling techniques are being considered for some locations. These scrambled signals necessarily cut into profits, consequently, inexpensive but effective measures similar to conventional cable-TV scrambling are desirable. Essentially, this arrangement consists of transmitting a precisely located interfering carrier within the video bandwidth, and removing that interfering carrier (for a monthly charge) with a special filter at the subscriber's location.

Each MDS subscriber is then provided with a receiving unit which downconverts 2.1 GHz to TV channel 2, 3 or 4 as required in the area. Descrambling filters connect between the converter and the television receiver. In addition to the monthly MDS charge, an installation fee and possibly a converter deposit are often charged (each lost or destroyed converter also cuts into MDS profits).

The question of whether an individual with a get-rich-quick gleam in his eye should consider joining the MDS game is debatable. The first thoughts give such a project appeal: invest in a TVRO ($4,000), a video-tape player and monthly tape-exchange service ($1,000), a 10-watt-output MDS transmitter and 2.1-GHz antenna ($2,000), and 50 or 60 receiving converters ($2,000). Sharp readers will notice that this $9,000 figure could be reduced to near $6,000 during startup of a station allowing the system to grow with income and acceptance.

In addition to financial considerations for MDS station equipment, another vitally important hurdle must be overcome before one can actually take to the airwaves: the MDS station must be licensed by the Federal Communications Commission, and permission to transmit television programming must be secured. The FCC doesn't grant a license to use the airwaves without justification: a meaningful purpose must be served in a constructive and positive manner, and the new service must not infringe on rights or fanchises of others also using the airwaves. The prospective microwave-TV enthusiast must do his homework in a number of areas before applying to the FCC for a license. In addition to polling the projected coverage area for outlining its audience and their requirements, a frequency/signal-strength/footprint study of the area must be conducted. This will ensure the new signal will peacefully co-exist with other franchised services while serving the intended area in a reliable manner. An additional market analysis considering future evolutions and their funding according to projected revenues at that time should also be considered, along with continued growth in the wake of possible future cable systems or public-broadcast satellite-TV networks. Fulfilling the previously outlined parameters, the individual might then file with the Federal Communications Commission and feel confident of results—providing a counterfile isn't also registered by another individual or a large corporation. Previously, some groups or individuals have made a habit of practically camping on the FCC's doorsteps merely for the purpose of filing on top of others and thus throwing applications into court battles. The initial applicant, possibly drained of extra funds from previous surveys and analysis, is then faced with additional cash outlays or watching his efforts go down the tube (pardon the expression). Fortunately, this blocking or counterfiling is usually confined to commercial television rather than microwave TV, but since that possibility exists, safeguarding initial efforts are a truly worthwhile consideration. Educational and religious groups are often exempt form the "counterfile syndrome." Indeed, commercial groups often assist these applicants in an effort to protect their own domain from competition by new sources. Assuming, however, the individual acquires a license and permission for microwave-TV (MDS) operation, the basic hurdles are behind him and the future is quite promising.

MDS-BAND EQUIPMENT

The equipment used for transmitting and receiving MDS television programming in the 2.1-GHz range is relatively similar to amateur gear used in the 2.3-GHz range. Indeed, many of the receiving antennas are identical for the two services. MDS equipment must necessarily be cost-effective; that is, it must do its required job at the least possible cost. Consequently, many of the leased MDS receive downconverters are less sensitive than similar amateur units (individual amateurs can usually afford to invest more "tweaking time" than can large companies).

MDS transmitters must be FCC type-approved. Although this requirement ensures top quality, it substantially increases unit costs. These transmitters are usually available from the many MDS/microwave equipment manufacturers advertising each month in electronic publications.

MDS receive downconverters take several shapes, but an in-

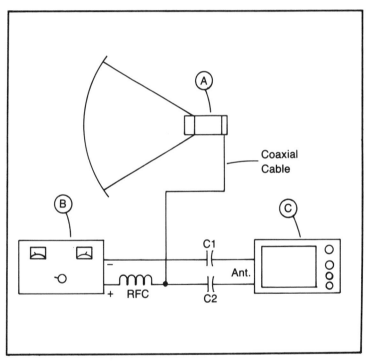

Fig. 12-7. Outline for a 2-GHz MDS setup as described in the text. Shown at (A) is a 2-pound coffee can with pickup antenna inside and downconverter board mounted on the rear; at (B) is a variable power supply, 7 to 12 volts regulated output; and at (C) a regular TV set tuned to the i-f from the downconverter—channel 3 or 4. Varying the voltage from the power supply tunes the downconverter to different signal frequencies. The rf choke isolates the i-f signal from the power supply, and the blocking capacitors (C1, C2) pass the i-f to the TV receiver while blocking the dc.

creasing number are following amateur-pioneered arrangements, such as that illustrated in Fig. 12-7. If this layout seems similar to those described in the previous 2.3-GHz amateur chapter, you're correct. The two regions, MDS and amateur 2.3 GHz, have quite a few aspects in common. Additional information on both satellite-TV reception and microwave TV (MDS) will be found in *Video Electronics Technology*, (TAB Book No. 1474), also by this author.

AMSAT

Radio Amateur Satellite Corporation

P.O. BOX 27, WASHINGTON, D.C. 20044

(202) 488-8649

AN INVITATION . . .
TO BECOME AN AMSAT LIFE MEMBER

For a contribution of $100 or more, you're invited to become an AMSAT Life Member. A bargain considering this is only 10 years' dues, and it's a tax-deductible donation. You may contribute in two or four installments if you prefer-- we only ask that you complete your payments by the end of the year. You can also charge your Life Membership to BankAmericard (VISA) or Master Charge.

<u>As a Life Member,</u>

● You receive the quarterly "AMSAT Newsletter" for life, now sent to Life Members by first class mail. If you're overseas, it's sent via airmail.

● We'll send you free, on request, AMSAT's W6PAJ orbital predictions book, which lists AMSAT-OSCAR satellite passes and operating schedules, annually.

● You'll receive other tokens of your support, such as an AMSAT-OSCAR pin, and a Life Membership certificate and card.

● Most importantly, your Life Membership helps sustain the amateur satellite program, making it possible for AMSAT to bring you newer and better satellites, serving as resources for the amateur community to use.

Won't <u>you</u> join us in Life Membership?

Fig. A-1. AMSAT life membership application.

Appendix A
AMSAT

AMSAT, the Radio Amateur Satellite Corporation, P.O. Box Washington, DC, 20044, was founded in 1969. The purposes objectives of this corporation are as follows:

A. To provide satellites that can be used for amateur ra communication and experimentation by suitably equip amateur radio stations throughout the world on a n discriminatory basis.

B. To encourage development of skills and the advanceme of specialized knowledge in the art and practice of amate radio communications and space science.

C. To foster international goodwill and cooperation throu joint experimentation and study, and through the w participation in these activities on a noncommercial ba by radio amateurs of the world.

D. To facilitate communications by means of amateur sa lites in times of emergency.

E. To encourage the more effective and expanded use of higher frequency amateur bands.

F. To disseminate scientific, technical and operational inf mation derived from such communications and experim tation, and to encourage publication of such informatio treatises, theses, trade publications, technical journals other public media.

Membership in AMSAT is open to all radio amateurs and ot interested persons. AMSAT encourages the participation of all terested individuals in its activities regardless of membership invites licensed amateur radio operators of all countries to engag radio transmissions to the satellite(s). Membership is possible in categories:

AMSAT
Radio Amateur Satellite Corporation
P.O. BOX 27, WASHINGTON, D.C. 20044
(202) 488-8649

TO: AMSAT Life Membership Chairman

FROM: _____
　　　　Name (print)　　　　　　Call

　　　_____
　　　Street address

　　　_____
　　　City, State (Country)　　ZIP/postal code

　　Yes, make me an AMSAT Life Member. I wish
to contribute $_____.

Method of payment:

　　Full payment enclosed ☐

　　Charge BankAmericard (VISA) ☐

　　Charge Master Charge ☐

_____
Charge card Account Number　　Expir. date

Payment in ☐ two or ☐ four installments.
Enclosed is $_____ as first installment.
I agree to contribute $_____ on each of
the following dates: _____
_____

_____　　　　　　　_____
　Date　　　　　　　　　Signature

Fig. A-1. (Continued.)

A. An interested individual may become a *Member* by filling out and returning the membership application, along with his dues payment (Fig. A-1).
B. A recognized group or organization interested in supporting AMSAT's goals and objectives and wishing to participate constructively in its activities may become a *Member Society* by completing and returning a member society application together with their dues payment. This class of membership was established to encourage interested groups to participate in AOSAT projects within the members society's own area of interest.

An annual financial contribution is requested from members and member societies in an effort to offset the costs of printing and mailing newsletters. This donation may be waived at the discretion of the Board of Directors. Donations are tax deductible.

Members of AMSAT are entitled to:
A. The opportunity to participate in the activities of AMSAT and to vote in the elections for the Board of Directors.
B. Receive newsletters and other information which may be generally distributed by AMSAT (Table A-1).
C. Be acknowledged as supporting the activities of AMSAT with a membership certificate or card.

Member Societies of AMSAT are entitled to:
A. Participate in AMSAT's acitivities.
B. Nominate two members per annum as candidates to the Board of Directors.
C. Receive newsletters and other information which may be distributed by AMSAT.
D. Be acknowledged as supporting AMSAT's activities with a Member Society card or certificate.

Since AMSAT's formation in 1969, this group has been responsible for several highly successful satellite projects. These projects include OSCAR 5, 6, 7, and 8. AMSAT is presently involved in the highly sophisticated Phase III satellite project (Fig. A-2) which will open new eras of UHF communications during late 1979 and the 1980's.

Table A-1. AMSAT Nets.

International Net	14,282 kHz	Sundays 1900 GMT
East Coast U.S. Net	3850 kHz	Wednesdays 0100 GMT
Mid U.S. Net	3850 kHz	Wednesdays 0200 GMT
West Coast U.S. Net	3850 kHz	Wednesdays 0300 GMT

AMSAT
Radio Amateur Satellite Corporation
P.O. BOX 27, WASHINGTON, D.C. 20044

HELP TO REVOLUTIONIZE AMATEUR RADIO COMMUNICATIONS!

Here's a unique opportunity for you to become a part of an amateur satellite project. How? By sponsoring solar cells and other portions of the new AMSAT Phase III satellites.

An exciting new phase in amateur radio is about to begin, one that will affect all of us. OSCAR satellites of the new AMSAT Phase III series will soon revolutionize long-distance amateur communications in the same manner that earth-bound repeaters have completely transformed local communications -- by dramatically increasing communications reliability while simultaneously reducing the cost and complexity of individual amateur stations. The first Phase III spacecraft, launch during 1983, will be available to most stations about 17 hours each day, and will make communications possible between stations separated by distances of up to 11,000 miles.

Operators interested in DX, rag-chewing, contests and traffic handling will find Phase III satellites as easy to use as their favorite band. In effect, each satellite will provide a new band comparable to 20 meters, a resource usable by hundreds of stations at a time.

But your help is needed to make Phase III a success. Hardware costs for the Phase III project are expected to be $250,000. (A commercial satellite of similar capability would cost nearly $10,000,000). You can help by sponsoring one or more components in the new satellites with tax-deductible donations ranging from $10 to $10,000.

Won't you please help? Please return the enclosed sponsorship form with your contribution today. We'll send you a certificate acknowledging the specific components you are sponsoring.

Thank you for your support!

Fig. A-2. AMSAT Phase III sponsorship form.

Radio Amateur Satellite Corporation
P.O. BOX 27, WASHINGTON, D.C. 20044

AMSAT-PHASE III SPONSORSHIP FORM

TO: AMSAT Phase III

FROM: Name_____ Call_____

Address_____

_____
City, State (Country)   ZIP

YES, I want to support AMSAT Phase III satellites. I wish to contribute $_____ in sponsorship of:

____ solar cells at $10 each
____ battery cells at $200 each
____ solar panels at $2,000 each
____ transponders at $5,000 each
____ rocket motor at $10,000

Method of payment:

☐ Check or money order enclosed

☐ Charge ☐ BankAmericard (VISA) or Master Charge ☐ Account Number: _____ Exp. date:_____

☐ Payment in ☐ two or ☐ four installments. Enclosed is $_____ as first installment. I agree to contribute $_____ on each of the following dates:

_____

Contributors of $1,000 or more will be honored by having their names inscribed on a plaque to be placed aboard the Phase III satellite for posterity. If you qualify, please indicate here how you would like your name and call listed:

_____

Sponsorship donations are tax-deductible.
Date:_____   Signature:_____

Fig. A-2. (Continued.)

Appendix B
Phonetic Alphabet

A	ALFA
B	BRAVO
C	CHARLIE
D	DELTA
E	ECHO
F	FOXTROT
G	GOLF
H	HOTEL
I	INDIA
J	JULIETTE
K	KILO
L	LIMA
M	MIKE
N	NOVEMBER
O	OSCAR
P	PAPA
Q	QUEBEC
R	ROMEO
S	SIERRA
T	TANGO
U	UNIFORM
V	VICTOR
W	WHISKEY
X	X-RAY
Y	YANKEE
Z	ZULU

Appendix C
International "Q" Signals

QRA	What is the name of your station?
QRB	How far are you from my station?
QRD	Where are you going and where are you from?
QRG	Will you tell me my exact frequency?
QRH	Does my frequency vary?
QRI	How is the tone of my transmission?
QRK	What is the readability of my signals?
QRL	Are you busy?
QRM	Are you being interfered with?
QRN	Are you troubled by static?
QRO	Shall I increase power?
QRP	Shall I decrease power?
QRQ	Shall I send faster?
QRR	Are you ready for automatic operation?
QRS	Shall I send more slowly?
QRT	Shall I stop sending?
QRU	Have you anything for me?
QRV	Are you ready?
QRW	Shall I inform _____ that you are calling him on _ kHz.
QRX	When will you call me again?
QRY	What is my turn?
QRZ	Who is calling me?
QSA	What is the strength of my signals?
QSB	Are my signals fading?
QSD	Is my keying defective?

QSG	Shall I send _____ telegrams at a time?
QSJ	What is the charge to be collected per word to _____ including your internal telegraph charge?
QSK	Can you hear me between your signals?
QSL	Can you acknowledge receipt?
QSM	Shall I repeat the last telegram which I sent you, or some previous telegram?
QSN	Did you hear me on _____ kHz?
QSO	Can you communicate with _____ direct or by relay?
QSP	Will you relay to _____ free of charge?
QSQ	Have you a doctor on board [or is ... (name of person) on board]?
QSU	Shall I send or replay on this frequency [or on _____ kHz] (with emission of class_____)?
QSV	Shall I send a series of V's on this frequency?
QSW	Will you send on this frequency?
QSX	Will you listen to _____ on _____ kHz?
QSY	Shall I change to transmission on another frequency?
QSZ	Shall I send each word or group more than once?
QTA	Shall I cancel telegram number _____ as if it had not been sent?
QTB	Do you agree with my counting of words?
QTC	How many telegrams have you to send?
QTE	What is my true bearing from you?
QTG	Will you send two dashes of ten seconds each followed by your call sign [on _____ kHz]?
QTH	What is your location?
QTI	What is your true track?
QTJ	What is your speed?
QTL	What is your true heading?
QTN	At what time did you leave _____ (place)?
QTO	Have you left dock (or port)? or Are you airborne?
QTP	Are you going to enter dock (or port)? or Are you going to land?

QTQ	Can you communicate with my station by means of the International Code of Signals?
QTR	What is the correct time?
QTS	Will you send your call sign for _____ minute(s) on _____ kHz so that your frequency may be measured?
QTU	What are the hours during which your station is open?
QTV	Shall I stand guard for you on the frequency of _____ kHz?
QTX	Will you keep your station open for further communication with me until further notice?
QUA	Have you news of _____ (call sign)?
QUB	Can you give me, in the following order, information concerning: visibility, height of clouds, direction and velocity (place of observation)?
QUC	What is the number of the last message you received from me?
QUD	Have you received the urgency signal, sent by _____ (call sign of mobile station)?
QUF	Have you received the distress signal sent by _____ (call sign of mobile station)?
QUG	Will you be forced to land?
QUH	Will you give me the present barometric pressure at sea level?

Appendix D

Great Circle Bearings (Beam Headings)

PREFIX	COUNTRY	BEARING, DEGREES ATLANTA, GA	DALLAS, TX	SACRAMENTO, CA
A5	BHUTAN, SIKKIM	6	355	332
AP	E. PAKISTAN	5	352	329
BV	FORMOSA	331	322	307
BY/C	CHINA	344	332	315
C6	BAHAMAS	108	108	96
CE	CHILE	167	156	140
CEØA	EASTER IS.	205	192	167
CEØZ	J. FERNANDEZ	173	163	146
CM, CO	CUBA	163	120	100
CN	MOROCCO	68	62	52
CP	BOLIVIA	160	144	125
CR5	SAO THOME	89	82	65
CR6	ANGOLA	93	86	66
CR7	MOZAMBIQUE	90	82	48
CR8	TIMOR	307	288	277
CR9	MACAO	340	327	306
CT1	PORTUGAL	62	56	43
CT2	AZORES	60	64	58
CT3	MADERIA	70	65	57
CX	URUGUAY	155	146	132
DJ, DL	GER. FED. REP.	40	39	30
DJ, DL	WEST BERLIN	42	36	27
DU	PHILIPPINES	325	314	298
EA	SPAIN	56	52	42
EA6	BALERIC IS.	57	51	42
EA8	CANARY IS.	72	69	49
EA9	CEUTA AND MELILLA	71	65	53
EA9	SP. MORROCO	78	57	52
EAØ	RIO MUNDI	63	77	46
EI	REP. IRELAND	84	42	35
EL	LIBERIA	92	86	72
EP	IRAN	36	25	25

PREFIX	COUNTRY	BEARING, DEGREES		
		ATLANTA, GA	DALLAS, TX	SACRAMENTO, CA
ET	ETHIOPIA	34	53	27
F	FRANCE	49	44	34
FB8Z	AMSTERDAM & ST. PAUL IS.	115	142	264
FB8X	KERGUELEN IS.	140	156	210
FC	CORSICA	52	46	35
FG7	GUADELOUPE	123	108	94
FH	MAYOTTE	79	70	31
FK8	NEW CALEDONIA	260	254	240
FM7	MARTINIQUE	123	110	97
FO8	CLIPPERTON IS.	129	206	154
FO8	TAHITI	241	232	209
FP8	ST. PIERRE & MIQUELON IS.	49	53	58
FR7	REUNION IS.	83	69	20
FS7	SAINT MARTIN IS.	122	107	94
FU8/YJ	NEW HEBRIDES IS.	272	266	352
FW8	WALLIS & FUTUNA	262	252	232
FY7, 8R	FRENCH GUIANA	125	110	100
G, GB	ENGLAND	43	40	32
GC, GU	GUERNSEY & DEPENDENCIES	48 42	44	33
GD	ISLE OF MAN	42	38	34
GI	NO. IRELAND	41	38	33
GM	SCOTLAND	42	39	32
GW	WALES	41	38	31
HA	HUNGARY	43	38	26
HB	SWITZERLAND	46	42	30
HE, HBØ	LIECHTENSTEIN	47	43	31
HC	ECUADOR	170	143	124
HC8	GALAPAGOES IS.	191	167	137
HH	HAITI	137	115	101
HI	DOMINICAN REPUBLIC	132	112	98
HK	COLOMBIA	159	138	115
HKØ	S. ANDRES & PROVIDENCIA IS.	171	135	122
HL, HM	KOREA (SOUTH)	332	325	311
HP	PANAMA	168	141	111
HR	HONDURAS	187	148	119
HS	THAILAND	351	335	312
HV	VATICAN	52	44	32
HZ	SAUDI ARABIA	46	36	10
II	ITALY	51	45	32
ISI	SARDINIA	53	48	35
ITI	SICILY	52	47	34
JA, KA	JAPAN	324	316	303
JT	MONGOLIA	353	343	330
JW	SVALBARD	12	11	7
JX	JAN MAYEN	24	21	18
JY	JORDAN	47	38	16
KB6, KH8	BAKER, HOWLAND, & AM. PHOENIX IS.	270	259	243
KC4	NAVASSA IS.	149	120	102

PREFIX	COUNTRY	BEARING, DEGREES		
		ATLANTA, GA	DALLAS, TX	SACRAMENTO, CA
KC6	CAROLINE IS.	298	290	275
KG4	GUANTANAMO BAY	141	117	99
KG6	GUAM	305	296	280
KG6R,S,T	MARIANAS IS.	312	307	286
KH6	HAWAII	280	273	252
KL7	ALASKA	330	326	340
KM6	MIDWAY IS.	294	290	375
KP4	PUERTO RICO	127	110	96
KS4	SWAN IS.	177	138	110
KS6	AM. SAMOA	258	249	232
KV4	VIRGIN IS.	124	108	97
KW6	WAKE IS.	296	289	276
KX6	MARSHALL IS.	287	280	265
LA	NORWAY	31	28	21
LU	ARGENTINA	162	153	138
LX	LUXEMBOURG	45	42	32
LZ	BULGARIA	44	38	22
OA	PERU	166	150	128
OD5	LEBANON	146	38	20
OE	AUSTRIA	167	41	28
OH	FINLAND	31	37	17
OHØ	ALAND IS.	30	36	16
OK	CZECHOSLOVAKIA	41	36	25
ON	BELGIUM	44	40	31
OX	GREENLAND	7	12	18
OY	FAROES	30	32	26
OZ	DENMARK	37	33	24
P29	NEW GUINEA	252	278	267
PA	NETHERLANDS	43	39	30
PJ	NETH. ANTILLES	144	122	104
PJ-M	SAINT MAARTEN	122	107	92
PY	BRAZIL	140	131	114
PY-Ø	TRINDADE	127	119	106
PZ	SURINAM	130	116	107
SL, SM	SWEDEN	32	30	21
SP	POLAND	37	34	23
ST	SUDAN	63	55	32
SU	EGYPT	51	43	23
SV	GREECE	50	32	24
SV	DODACANESE IS.	49	35	26
TA	TURKEY	43	35	18
TF	ICELAND	30	30	28
TG	GUATEMALA	199	159	121
TI	COSTA RICA	178	148	119
TJ	CAMERON	83	75	57
TL	CEN. AF. REP.	76	68	46
TN	CONGO REP.	88	80	60
TR	GABON. REP.	87	79	59
TT	CHAD REP.	71	65	43
TU	IVORY COAST	87	79	63

PREFIX	COUNTRY	BEARING, DEGREES		
		ATLANTA, GA	DALLAS, TX	SACRAMENTO, CA
TZ	MALI REP	86	78	63
UA, UV, UW, U UA, UW	EUROPEAN USSR	30	24	16
9, Ø	ASIATIC USSR	3	357	9
UB5, UT5	UKRAIN	27	27	18
UC2	WHITE RUSSIA	35	30	18
UF6	GEORGIA	35	27	10
UG6	ARMENIA	36	26	9
UH8	TURKOMAN	25	16	358
UI8	UZBEK	24	16	356
UJ8	TADZHIK	18	8	345
UL7	KAZAKH	18	10	352
UM8	KIRGHIZ	17	8	347
UNI	KARELO-FINN REPUBLIC	25	21	11
UO5	MOLDAVIA	40	34	18
UP2	LITHUANIA	33	28	19
UQ2	LATVIA	32	27	16
UR2	ESTONIA	31	28	18
VK	AUSTRALIA	255	250	243
VK9	NORFOLK IS.	252	247	235
VKØ	HEARD IS.	176	197	226
VKØ	MACQUARIE IS.	224	223	220
VP5	TURKS & CAICOS IS.	127	108	96
VP8, LU-Z	S. GEORGIA IS.	157	150	141
VP8, LU-Z	S. FALKLANDS	159	151	147
VP8, LU-Z	S. ORKNEYS	157	148	140
VP9	BERMUDA	89	83	79
VQ8	CHAGOES IS.	70	55	356
VQ8	MAURITIUS IS.	74	58	346
VQ8	RODRIGUEZ IS.	72	56	347
VR1	GILBERT & ELLICE AND OCEAN IS.	272	262	249
VR2	FIJI IS.	258	252	237
VR4	SOLOMON IS.	275	269	255
VR6	PITCAIRN IS.	220	212	188
VU2	ANDAMAN & NICOBAR IS.	3	245	320
VU2	INDIA	25	11	343
VU2	LACCADIVE IS.	27	12	340
XE, XF	MEXICO	227	185	126
XE, XF	REVILLA GIGEDO	246	226	154
XW8	LAOS	351	334	310
YA	AFGHANISTAN	23	12	350
Y1	IRAQ	42	34	16
YK	SYRIA	41	36	17
YJ	NEW HEBRIDES	266	259	246
YN	NICARAGUA	180	148	115
YO	RUMANIA	43	36	24
YS	SALVADOR	193	155	120
YU	YUGOSLAVIA	43	38	26
YV, 4M	VENEZUELA	140	122	105

		BEARING, DEGREES		
PREFIX	COUNTRY	ATLANTA, GA	DALLAS, TX	SACRAMENTO, CA
ZA	ALBANIA	48	42	46
ZD7	ST. HELENA IS.	108	101	85
ZD8	ASCENSION IS.	110	103	87
ZD9	TR. D. CUNHA & GOUGH IS.	130	125	113
ZE	ZIMBABWE	92	86	58
ZK1	COOK IS.	249	242	220
ZK1	MANIHIKI IS.	250	241	223
ZK2, C21	NIUE IS.	251	244	227
ZL	NEW ZEALAND	238	233	223
ZP	PARAGUAY	152	140	125
ZS1,2,4 5 & 6	SOUTH AFRICA	102	97	74
ZS3	S. W. AFRICA	105	98	79
IS	SPRATLY ISLAND	330	322	307
3A	MONACO	50	45	34
3V8	TUNISIA	57	52	38
3W8	VIET NAM	340	312	291
4S7	CEYLON	21	40	331
4U	GENEVA	49	44	33
4W	YEMAN	54	43	17
4X	ISRAEL	47	39	20
4X1	ISRAEL	40	40	??
5A	LIBYA	59	52	38
5B4, ZC4	CYPRUS	45	39	21
5H3	TANZANIA	79	70	39
5N2	NIGERIA	85	78	61
5R8	MALAGASY REP.	84	75	30
5T	MAURITANIA REP.	86	78	63
5U7	NIGER REP.	77	70	54
5V	TOGO REP.	87	80	65
5W1	W. SAMOA	257	250	232
5X5	UGANDA	73	64	38
5Z4	KENYA	73	65	37
6O	SOMALI REP.	62	50	20
6W8	SENEGAL REP.	89	82	70
7G1	REP. OF GUINEA	92	86	67
7X	ALGERIA	58	53	41
8Z4, 8Z5	NEUTRAL ZONE	42	32	15
9H1	MALTA	55	49	35
9J2	ZAMBIA	42	83	58
9K2	KUWAIT	348	32	16
9L1	SIERRA LEONE	92	85	71
9M2	WEST MALAYSIA	349	329	306
9M6, 8	EAST MALAYSIA	330	314	296
9V1	SINGAPORE	344	325	300
9N1	NEPAL	10	358	332
9Q5	REP. OF ZAIRE	83	75	51
9U5	BURUNDI	79	71	44
9X5	RWANDA	78	69	42
9Y4(VP4)	TRINIDAD & TOBAGO	129	115	100

Appendix E
International Prefixes

PREFIX	COUNTRY
AP	PAKISTAN
A2C	BOTSWANA
A35	TONGA
A4X	SULTANATE OF OMAN
A51	BHUTAN
A6X	UNITED ARAB EMIRATES
A7X	QATAR
A9X	BAHRAIN
BV	TAIWAN
BY	PEOPLES REPUBLIC OF CHINA
CE	CHILE
CE9AA-AM	CHILEAN ANTARCTICA
CE9AN-AZ	SOUTH SHETLAND IS.
CE0A	EASTER IS.
CE0X	SAN FELIX IS.
CE0Z	JUAN FERNANDEZ IS.
CM	CUBA
CN	MOROCCO
CO	CUBA
CP	BOLIVIA
CR3	GUINEA BISSAU
CR9	MACAO
CT1	PORTUGAL
CT2	AZORES IS.
CT3	MADEIRA IS.
CX	URUGUAY
C21	REPUBLIC OF NAURU
C31	ANDORRA
C5A	GAMBIA
C6A	BAHAMA IS.
C9M	MOZAMBIQUE
DU	PHILIPPINES
DA-DL	FED REP OF WEST GERMANY
DM	EAST GERMAN DEMOCRATIC REP.
D2A,D3	PHILIPPINES
D2A	ANGOLA
D4	REP OF CAPE VERDE
D6	STATE OF COMORO

EA	SPAIN
EA6	BALEARIC IS.
EA8	CANARY IS.
EA9	SPANISH SAHARA, CEUTA Y MELILLA
EI	IRELAND
EL,5L	LIBERIA
EP	IRAN
ET3	ETHIOPIA
F	FRANCE
FB8W	CROZET IS.
FB8X	KERGUELEN IS.
FB8Y	ANTARCTICA
FB8Z	AMSTERDAM & ST. PAUL IS.
FC	CORSICA
FG7	GUADELOUPE
FH8	MAYOTTE
FK8	NEW CALEDONIA
FM7	MARTINIQUE
FO8	FRENCH POLYNESIA, CLIPPERTON IS.
FP8	ST. PIERRE & MIQUELON IS.
FR7	GLORIOSO, JUAN DE NOVA, REUNION, TROMELIN IS.
FS7	ST. MARTIN IS.
FW8	WALLIS & FUTUNA IS.
FY7	FRENCH GUIANA
G	ENGLAND
GD	ISLE OF MAN
GI	NORTHERN IRELAND
GJ	JERSEY
GM	SCOTLAND
GU	BAILIWICK OF GUERNSEY
GW	WALES
HA,HG	HUNGARY
HB	SWITZERLAND
HB0	LIECHTENSTEIN
HC	ECUADOR
HC8	GALAPAGOS IS.
HG	HUNGARY
HH	HAITI
HI	DOMINICAN REPUBLIC
HK	COLOMBIA
HK0	BAJO NUEVO, MALPELO, SAN ANDRES & PROVIDENCIA IS.
HK0	SERRANA BANK & RONCADOR CAY
HM,HL9	KOREA
HP	PANAMA
HR	HONDURAS
HS	THAILAND
HV	VATICAN CITY
HZ,7Z	SAUDI ARABIA
H4	SOLOMON IS.
I, IW	ITALY
IC	CAPRI & ISCHIA IS.

IG9	LAMPEDUSA IS.
IH	PANTELLERIA IS.
IM	MADDALENA IS.
IS	SARDINIA
IT	SICILY
JA,JE-JJ,JR	JAPAN
JD1	OGASAWARA,MINAMI-TORI-SHIMA IS.
JR6	OKINAWA (RYUKYU IS.)
JT1	MONGOLIA
JW	SVALBARD IS.
JX	JAN MAYEN IS.
JY	JORDAN
J2	REP OF DJIBOUTI
J3	GRENADA & DEPENDENCIES
J6,VP2L	ST LUCIA
K	UNITED STATES OF AMERICA
KA	U.S.PERSONNEL IN JAPAN
KB6	BAKER,CANTON,ENDERBURY, HOWLAND & PHOENIX IS.
J7,VP2L	DOMINICA
KC4	NAVASSA IS.
KC4AA,KC4US	ANTARCTICA
KC6	CAROLINE IS.
KG4	GUANTANAMO BAY
KG6	MARIANA IS.
KG6	GUAM
KG6R	ROTA
KG6S	SAIPAN
KH6	HAWAII & KURE IS.
KJ6	JOHNSTON IS.
KL7	ALASKA
KM6	MIDWAY IS.
KP4	PUERTO RICO
KP6	JARVIS & PALMYRA IS.
KS6	AMERICAN SAMOA
KV4	VIRGIN IS.
KW6	WAKE IS.
KX6	MARSHALL IS.
LA-LJ	NORWAY
LU	ARGENTINA
LU-Z	ANTARCTICA
LX	LUXEMBOURG
LZ	BULGARIA
M1	SAN MARINO
N	UNITED STATES OF AMERICA
OA	PERU
OD5	LEBANON
OE	AUSTRIA
OH	FINLAND
OH0	ALAND IS.
OJ0, OH0M	MARKET REEF

OK,OL	CZECHOSLOVAKIA
ON	BELGIUM
OR	ANTARCTICA
OX	GREENLAND
OY	FAEROES IS.
OZ	DENMARK
PA-PI	NETHERLANDS
PJ	NETHERLANDS ANTILLES
PJ2,9	CURACAO
PJ3,9	ARUBA
PJ4,9	BONAIRE
PJ5,8	ST. EUSTATIUS
PJ6,8	SABA IS.
PJ7,8	SINT MAARTEN
PP-PY	BRAZIL
PY0	FERNANDO DE NORONHA IS.
PY0	ST. PETER & ST. PAUL'S ROCKS
PY0	TRINIDADE & MARTIM VAZ IS.
PZ	SURINAM
P29	PAPUA NEW GUINEA
SJ-SM	SWEDEN
SP	POLAND
ST	SUDAN
SU	EGYPT
SV	CRETE, GREECE
SV5	DODECANESE IS.
S2,S3	BANGLADESH
S7	SEYCHELLES IS.
S8	TRANSKEI
S9	SAO TOME & PRINCIPE IS.
T2	TUVALU ISLANDS
TA,TC	TURKEY
TF	ICELAND
TG	GUATEMALA
TI	COSTA RICA
TI9	COCOS IS.
TJ	CAMEROON
TL8	CENTRAL AFRICAN REPUBLIC
TN8	REPUBLIC OF CONGO
TR8	GABON REPUBLIC
TT8	REPUBLIC OF CHAD
TU2	IVORY COAST
TY	PEOPLES REPUBLIC OF BENIN
TZ	MALI REPUBLIC
UA1,2,3,4,6	EUROPEAN RUSSIAN SOVIET FEDERATED SOCIALIST REPUBLIC
UA9,0	ASIATIC RUSSIAN S.F.S.R.
UA1	FRANZ JOSEF LAND
UA2	KALININGRADSK
UB5	UKRAINIAN S.S.R.
UC2	WHITE RUSSIAN S.S.R.
UD6	AZERBAIDZHAN S.S.R.
UF6	GEORGIAN S.S.R.
UG6	ARMENIAN S.S.R.

Prefix	Entity
UH8	TURKMEN S.S.R.
UI8	UZBEK S.S.R.
UJ8	TADZHIK S.S.R.
UL7	KAZAKH S.S.R.
UM8	KIRGHIZ S.S.R.
UN1	KARELO-FINNISH S.S.R.
UO5	MOLDAVIAN S.S.R.
UP2	LITHUANIAN S.S.R.
UQ2	LATVIAN S.S.R.
UR2	ESTONIAN S.S.R.
VE	CANADA
VK	AUSTRALIA
VK2	LORD HOWE IS.
VK9N	NORFOLK IS.
VK9X	CHRISTMAS IS.
VK9V	COCOS IS.
VK9Z	WILLIS IS.
VK0	ANTARCTICA
VO1	NEWFOUNDLAND
VO2	LABRADOR
VP1	BELIZE
VP2	LEEWARD & WINDWARD IS.
VP2A	ANTIGUA, BARBUDA
VP2O,J7	DOMINICA
VP2E	ANGUILLA
VP2K	ST KITTS, NEVIS
VP2L,J6	ST LUCIA
VP2M	MONSERRAT
VP2S	ST VINCENT & DEPENDENCIES
VP2V	BRITISH VIRGIN IS.
VP5	TURKS & CAICOS IS.
VP8	FALKLAND, S.GEORGIA, S.ORKNEY, S.SANDWICH, S.SHETLAND IS., GRAHAM LAND
VP9	BERMUDA IS.
VQ9	CHAGOS
VR1	BRITISH PHOENIX, GILBERT & OCEAN IS.
VR3	NORTHERN LINE IS.
VR6	PITCAIRN IS.
VR7	CENTRAL & SOUTHERN LINE IS.
VS5	BRUNEI
VS6	HONG KONG
VU2	INDIA
VU7	ANDAMAN & NICOBAR IS.
VU7	LACCADIVE IS.
W	UNITED STATES OF AMERICA
XE,XF	MEXICO
XF4	REVILLA GIGEDO IS.
XT2	VOLTAIC REPUBLIC
XU	CAMBODIA/KHMER REPUBLIC
XV5	VIETNAM
XW8	LAOS
XZ	BURMA

Prefix	Country
YA	AFGHANISTAN
YB-YD	INDONESIA,TIMOR IS.
YI	IRAQ
YJ	NEW HEBRIDES
YK	SYRIA
YN	NICARAGUA
YO	ROMANIA
YS	EL SALVADOR
YU	YUGOSLAVIA
YV	VENEZUELA
YV0	AVES IS.
ZA	ALBANIA
ZB2	GIBRALTAR
ZD7	SAINT HELENA IS.
ZD8	ASCENSION IS.
ZD9	TRISTAN DA CUNHA & GOUGH IS.
ZE	ZIMBABWE
ZF1	CAYMAN IS.
ZK1	COOK & MANIHIKI IS.
ZK2	NIUE IS.
ZL	NEW ZEALAND & AUCKLAND,CAMPBELL,CHATHAM,KERMADEC IS.
ZL5	ANTARCTICA
ZM7	TOKELAU IS.
ZP	PARAGUAY
ZR,ZS1,2,4,5,6	REPUBLIC OF SOUTH AFRICA
ZS1ANT	ANTARCTICA
ZS2	PRINCE EDWARD & MARION IS.
ZR,ZS3	SOUTHWEST AFRICA (NAMIBIA)
3A	MONACO
3B6	AGALEGA IS.
3B7	ST. BRANDON IS.
3B8	MAURITIUS IS.
3B9	RODRIGUEZ IS.
3C	EQUATORIAL GUINEA
3D2	FIJI IS.
3D6	SWAZILAND
3V8	TUNISIA
3X	REPUBLIC OF GUINEA
3Y	BOUVET IS.
4K1	ANTARCTICA
4S7	SRI LANKA
4U	UNITED NATIONS,GENEVA
4W	YEMEN
4X4,4Z4	ISRAEL
5A	LIBYAN ARAB REPUBLIC
5B4,ZC4	CYPRUS
5HI	ZANZIBAR,TANZANIA
5H3	TANZANIA
5L	LIBERIA
5N2	NIGERIA
5R8	MALAGASY REPUBLIC
5T5	MAURITANIA

5U7	NIGER
5V	TOGO
5W1	WESTERN SAMOA
5X5	UGANDA
5Z4	KENYA
6O	SOMALI REPUBLIC
6W8	SENEGAL REPUBLIC
6Y5	JAMAICA
7O	SOUTH YEMEN & KAMARAN IS
7P8	LESOTHO
7Q7	MALAWI
7X	ALGERIA
7Z	SAUDI ARABIA
8J	ANTARCTICA
8P6	BARBADOS IS.
8Q6	MALDIVE IS.
8R	GUYANA
8Z4	SAUDI ARABIA/IRAQ NEUTRAL ZONE
9G1	GHANA
9H1,5	MALTA
9H4	GOZO (MALTA)
9I,9J	ZAMBIA
9K2	KUWAIT
9L1	SIERRA LEONE
9M2	WEST MALAYSIA
9M6	SABAH
9M8	SARAWAK
9N1	NEPAL
9Q5	REPUBLIC OF ZAIRE
9U5	BURUNDI
9V1	REPUBLIC OF SINGAPORE
9X	RWANDA
9Y4	TRINIDAD & TOBAGO IS.

COUNTRY	*PREFIX*
AFGHANISTAN	YA
AGALEGA IS	3B6
ALAND IS	OH0
ALASKA	KL7
ALBANIA	ZA
ALGERIA	7X
AMSTERDAM & ST. PAUL IS	FB8Z
ANDAMAN & NICOBAR IS	VU7
ANGOLA	D2A
ANTARCTICA	CE9AA-AM,FBBY,KC4,LU-Z, UA1,VK0,VP8,ZL5,ZS1ANT,3Y,4K1,8J
ARGENTINA	LU
ARUBA	PJ3,9
ASCENSION IS	ZD8
AUCKLAND & CAMPBELL IS	ZL
AUSTRALIA	VK
AUSTRIA	OE

AVES IS	YV0
AZDRES IS	CT2
BAHAMA IS	C6A
BAHRAIN IS	A9X
BAJO NUEVO IS	HK0
BAKER IS	KB6
BALEARIC IS	EA6
BANGLADESH	S2,S3
BARBADOS IS	8P6
BELGIUM	ON
BELIZE	VP1
BENIN, PEOPLE'S REP OF	TY
BERMUDA IS	VP9
BHUTAN	A51
BOLIVIA	CP
BONAIRE	PJ4,9
BOTSWANA	A2C
BOUVET IS	3Y
BRAZIL	PP-PY
BRITISH PHOENIX IS	VR1
BRUNEI	VS5
BULGARIA	LZ
BURMA	XZ
BURUNDI	9U5
CAMBODIA/KHMER REP	XU
CAMEROON	TJ
CANADA	VE
CANARY IS	EA8
CANTON IS	KB6
CAPE VERDE, REP OF	D4
CAPRI IS	IC
CAROLINE IS	KC6
CAYMAN IS	ZF1
CENTRAL AFRICAN REPUBLIC	TL8
CEUTA Y MELILLA, SPANISH	EA9
CHAD REPUBLIC	TT8
CHAGOS IS	VQ9
CHATHAM IS	ZL
CHILE	CE
CHINA, PEOPLES REP OF	BY
CHRISTMAS IS	VK9X
CLIPPERTON IS	FO8
COCOS IS	TI9
COCOS (KEELING) IS	VK9Y
COLUMBIA	HK
COMORO, STATE OF	D6
CONGO, REPUBLIC OF	TN8
COOK IS	ZK1
CORSICA	FC
COSTA RICA	TI
CRETE, GREECE	SV
CROZET IS	FB8W
CUBA	CM,CO
CURACAO	PJ2,9

CYPRUS	5B4,ZC4
CZECHOSLOVAKIA	OK,OL
DENMARK	OZ
DJIBOUTI,REP OF	J2
DODECANESE IS	SV5
DOMINICAN REPUBLIC	HI
EASTER IS	CE0A
ECUADOR	HC
EGYPT	SU
EL SALVADOR	YS
ENDERBURY IS	KB6
ENGLAND	G
ETHIOPIA	ET3
FAEROES IS	OY
FALKLAND IS	VP8
FERNANDO DE NORONHA	PY0
FIJI IS	3D2
FINLAND	OH
FRANCE	F
FRENCH POLYNESIA	FD8
GABON REPUBLIC	TR8
GALAPAGOS IS	HC8
GAMBIA	C5A
GERMANY, FED. REP. (WEST)	DA-DL
GERMAN DEM. REP. EAST	DM
GHANA	9G1
GIBRALTAR	ZB2
GILBERT & OCEAN IS	VRI
GLORIOSO IS	FR7
GOUGH IS	ZD9
GOZO (MALTA)	9H4
GRAHAM LAND	VP8,LU-Z
GREECE	SV
GREENLAND	OX
GRENADA	J3
GUADELOUPE	FG7
GUAM IS	KG6
GUANTANAMO BAY	KG4
GUATEMALA	TG
GUERNSEY,BAILIWICK OF	GU
GUIANA, FRENCH	FY7
GUINEA BISSAU	CR3
GUINEA, EQUATORIAL	3C
GUINEA, REPUBLIC OF	3X
GUYANA	8R
HAITI	HH
HAWAII	KH6
HEARD IS	VK0
HONDURAS	HR
HONG KONG	VS6
HOWLAND IS	KB6
HUNGARY	HA, HG

ICELAND	TF
INDIA	VU2
INDONESIA	YB-YD
IRAN	EP
IRAQ	YI
IRELAND	EI
ISCHIA	IC
ISLE OF MAN	GD
ISRAEL	4X4,4Z4
ITALY	I,IW
IVORY COAST	TU2
JAMAICA	6Y5
JAN MAYEN IS	JX
JAPAN	JA-JR
JAPAN, U.S.PERSONNEL IN	KA
JARVIS IS	KP6
JERSEY	GJ
JOHNSTON IS	KJ6
JORDON	JY
JUAN DE NOVA IS	FR7
JUAN FERNANDEZ IS	CE0X
KAMARAN IS	7O
KENYA	5Z4
KERGUELEN IS	FB8X
DERMADEC IS	ZL
KOREA	HM,HL9
KURE IS	KH6
KUWAIT	9K2
LABRADOR	VO2
LACCADIVE IS	VU7
LAMPEDUSA IS	IG9
LAOS	XW8
LEBANON	OD5
LEEWARD IS.	
ANGUILLA	VP2E
ANTIGUA, BARBUDA	VP2A
BRITISH VIRGIN IS	VP2V
MONTSERRAT	VP2M
ST. KITTS, NEVIS	VP2K
LESOTHO	7P8
LIBERIA	EL,5L
LIBYAN ARAB REPUBLIC	5A
LIECHTENSTEIN	HB0
LINE IS, NORTHERN	VR3
LINE IS,CENTRAL & SOUTHERN	VR7
LORD HOWE IS	VK2
LUXEMBOURG	LX
MACAO	CR9
MACQUARIE IS	VK0
MADDALENA IS	IM
MADEIRA IS	CT3
MALAGASY REPUBLIC	5R8
MALAWI	7Q7

MALAYSIA, EAST	9M6,8
MALAYSIA, WEST	9M2
MALDIVE IS	8Q6
MALI REPUBLIC	TZ
MALPELO IS	HK0
MALTA	9H1,5
MANIHIKI IS	ZK1
MARIANA IS	KG6
MARKET REEF	OJ0,OH0M
MARSHALL IS	KX6
MARTINIQUE	FM7
MAURITANIA	5T5
MAURITIUS IS	3B8
MAYOTTE	FH8
MEXICO	XE,XF
MIDWAY IS	KM6
MINAMI-TORI-SHIMA IS	JD1
MONACO	3A
MONGOLIA	JT1
MOROCCO	ON
MOZAMBIQUE	O9M
NAURU, REPUBLIC OF	C21
NAVASSA IS	KC4
NEPAL	9N1
NETHERLANDS	PA-PI
NETHERLANDS ANTILLES	PJ
NEW CALEDONIA	FK8
NEWFOUNDLAND	VO1
NEW HEBRIDES	YJ
NEW ZEALAND	ZL
NICARAGUA	YN
NIGER	5U7
NIGERIA	5N2
NIUE IS	ZK2
NORFOLK IS	VK9N
NORTHERN IRELAND	GI
NORWAY	LA-LJ
OGASAWARA IS	JDI
OKINAWA (RYUKYU IS.)	JR6
OKINO-TORI-SHIMA	7J1
OMAN, SULTANATE OF	A4K
PAKISTAN	AP
PALMYRA IS	KP6
PANAMA	HP
PANTELLERIA IS	IH
PAPUA NEW GUINEA	P29
PARAGUAY	ZP
PERU	OA
PHILLIPINES	DU
PHOENIX IS	K36
PITCAIRN IS	VR6
POLAND	SP
PORTUGAL	CT1
PRINCE EDWARD & MARION IS.	Z52
PRINCIPE & SAO TOME IS	S9

PUERTO RICO	KP4
QATAR	A7K
REUNION IS	FR7
REVILLA GIGEDO IS	XF4
ZIMBABWE	ZE
RODRIGUEZ IS	3B9
RUMANIA	YO
ROTA	KG6R
RWANDA	9X
SABAH	9M6
SABA IS	PJ6,8
ST. BRANDON IS	3B7
ST. EUSTATIUS	PJ5,8
SAINT HELENA IS	ZD7
SAINT MARTIN IS	FS7
ST. PETER & ST. PAULS ROCKS	PY0
ST. PIERRE & MIQUELON IS	FP8
SAIPAN	KG6S
SAMOA, AMERICAN	KS6
SAMOA, WESTERN	5W1
SAN ANDRES & PROVIDENCIA	HK0
SAN FELIX IS	CE0X
SAN MARINO	M1
SARAWAK	9M8
SARDINIA	IS
SAUDI ARABIA	HZ, 7Z
SAUDI ARABIA/IRAQ NEUTRAL ZONE	8Z4
SCOTLAND	GM
SENEGAL REPUBLIC	6W8
SERRANA BK & RONCADOR CAY	8K0
SEYCHELLES IS	S79
SICILY	IT
SIERRA LEONE	9L1
SINGAPORE, REP OF	9V1
SINT MAARTEN	PJ7,8
SOLOMON IS	H4
SOMALI REPUBLIC	60
SOUTH AFRICA,REP OF	ZR,AZQ,2,4-6
SOUTH GEORGIA IS	VP8
SOUTH ORKNEY IS	QZ, 2, VP8
SOUTH SANDWICH IS	QZ, 2, VP8
S.SHETLAND IS	CE9AN-AZ,QZ-Z,VP8
SOUTHWEST AFRICA(NAMIBIA)	ZR3,ZS3
SOUTH YEMAN	70
SOVIET UNION:	
EUROPEAN RUSSIA SOVIET FEDERATED SOCIALIST REPUBLIC	UA1,2,3,4,6
ASIATIC RUSSIAN SFSR	UA9,0
ARMENIAN S.S.R	UG6
AZERBAIDZHAN S.S.R	UD6
ESTONIAN S.S.R	UR2
FRANZ JOSEF LAND	UA1
GEORGIAN S.S.R	UF6
KALININGRADSK	UA2

KARELO-FINNISH S.S.R	UN1
KAZAKH S.S.R	UL7
KIRGHIZ S.S.R	UMB
LATVIAN S.S.R	UQ2
LITHUANIAN S.S.R	UP2
MOLDAVIAN S.S.R	UO5
TADZHIK S.S.R	UJ8
TURKMEN S.S.R	UH8
UKRANIAN S.S.R	UB5
UZBEK S.S.R	UI8
WHITE RUSSIAN S.S.R	UC2
SPAIN	EA
SPANISH SAHARA	EA9
SRI LANKA	4S7
SUDAN	ST
SURINAM	P2
SVALBARD IS	JW
SWAN IS	HRO
SWAZILAND	3D6
SWEDEN	SJ-SM
SWITZERLAND	HB
SYRIA	YK
TAIWAN	BV
TANZANIA	5H3
THAILAND	HS
TIBET	AC4
TIMOR PORTUGUESE	YB-YD
TOGO	5V
TOKELAU IS	2M7
TUNGA	A35
TRANSKEI	S8
TRINIDADE & MARTIM VAZ IS	PYO
TRINIDAD & TOBAGO IS	9Y4
TRISTAN DE CUNHA	ZD9
TROMELIN IS	FR7
TUNISIA	3V8
TURKEY	TA,TC
TURKS & CAICOS IS	VP5
TUSCAN ARCHIPELAGO	IA
TUVALU IS	T2
UGANDA	5X5
UNITED ARAB EMIRATES	A6X
UNITED NATIONS, GENEVA	4U
UNITED STATES OF AMERICA	N,W,K
URUGUAY	CX
VATICAN CITY	HV
VENEZULA	YV
VIETNAM	XV5
VIRGIN IS	KV4
VOLTAIC REPUBLIC	XT2
WAKE IS	KW6
WALES	GW
WALLIS & FUTUNA IS	FW8
WILLIS IS	VK9Z

WINDWARD IS:
    DOMINICA J73
    ST. LUCIA J6
    ST. VINCIENT & DEPENDENCIES J8
    GRENADA J3
YEMEN 4W
YUGOSLAVIA .... YU

ZAIRE, REPUBLIC OF 9Q5
ZAMBIA 9I,9J
ZANZIBAR (TANZANIA) 5H1

Index

A
Advanced Receiver Research, 63
afc technique, 40
alarms
　intruder/motion-detector, 131
alphabet
　phonetic, 159
amateur band
　10 cm, 10
　13 cm, 10
　15 mm, 11
　23 cm, 9
　5 cm, 10
amateur fast-scan TV, 31
amateur RADAR, 128
amateur satellite program, 14
amateur system
　2.3 GHz, 52
amateur weather RADAR, 134
AMSAT, 153
antenna
　cigar, 59
　funnel, 61
　parabolic dish, 61, 113
　resonant-cavity mounted, 60
　13 cm, 34
antennas for 2.3 GHz, 58

B
batteries
　nickel-cadmium, 102

C
capacitance
　interelectrode, 19
capacitors, 96
circuits
　13 cm, 34
communications equipment
　1.2 GHz, 42
　2.3 GHz, 51
communications equipment for 10 GHz, 63

communications setup
　10 GHz, 66
computer communications, 85
CW, 42

D
digital automatic frequency control, 39
Doppler frequency, 130
DX
　23 cm, 48

E
earth-moon-earth path, 12
electromagnetic heating effects on human tissue, 116
electromagnetic spectrum, 1
EME, 12, 29
equipment designs
　3 cm, 37

F
fast-scan repeater, 2
fast-scan TV, 120
Federal Communications Commission, 150
FM, 8, 42

G
GaAsFET, 145
great circle bearings, 163
Gunn diode, 33
Gunn diode theory, 27
Gunnplexer, 32, 38, 65, 67
　10 GHz, 122
　phase-locked, 78

H
Hertz
　Heinrich, 3
home-computer linking, 55

I
interelectrode capacitance, 19, 20

international prefixes, 168
international "Q" signals, 160
intruder alarms, 128

K
klystron operation, 23

L
LNA, 145

M
magnetron operation, 25
Marconi
  Guglielmo, 3
MDS, 9, 29, 141, 146
  operational concepts of, 149
MDS equipment, 150
Microwave Associates, 63
microwave bands, 29
microwave electronic theory, 17
microwave equipment
  10 GHz, 6
microwave frequency allocations, 8
microwave interfacing, 119
microwave linking of home computers, 126
microwave network, 5
microwave networking, 85
microwave spectrum, 1, 7
microwave system expansion, 115
microwave system operation, 111
microwave system power supplies, 95
microwave system tuning, 111
multimode communications, 86
multipoint distribution systems, 147

N
network
  microwave, 5
networking
  microwave, 85

O
OSCAR, 8, 14, 29, 42
OSCAR 8, 15
OSCAR 9, 15
OSCAR phase III, 15
OSCAR phase IV geostationary satellites, 87
oscillator
  tuned-grid tuned-plate, 19

P
parabolic dish antenna, 113
phase-locked 10 GHz setup, 79

phase-locked Gunnplexer, 78
phonetic alphabet, 159
power sources
  natural, 102
power supplies, 95
power supply
  general purpose, 97
  pick-a-volt, 100
  safe-stop, 99
preamplifier
  high-performance receiving, 117

Q
QRP, 42
  2.3 GHz, 56

R
RADAR, 128
  10 GHz, 132
  amateur weather, 134
  types of, 129
regulators, 96
repeater
  fast-scan, 2
repeater system
  portable, 47
resonant cavities, 24
resonant-cavity mounted antenna, 60
RTTY, 2

S
safety considerations, 114
satellite-TV reception, 143
satellites, 9
  television broadcasting, 142
scan converting relays, 124
skin effect, 18
solar energy systems, 108
SSB, 8, 42
strip-line tuned circuit, 22

T
television broadcasting satellites, 142
TGTP oscillator, 18
transformers, 95
transmitter
  2.3 GHz, 57
  low-power 2.3 GHz, 59
tuned-grid tuned-plate oscillator, 19
TVRO, 141

W
water power systems, 103
wind power systems, 103

183

Other Bestsellers From TAB

☐ **ENCYCLOPEDIA OF ELECTRONICS**

Here are more than 3,000 complete articles covering many more thousands of electronics terms and applications. A must-have resource for anyone involved in any area of electronics or communications practice. From basic electronics or communications practice. From basic electronics terms to state-of-the-art digital electronics theory and applications . . . from microcomputers and laser technology to amateur radio and satellite TV, you can count on finding the information you need! 1,024 pp., 1,300 illus. 8 1/2″ × 11″.

Hard $60.00   Book No. 2000

☐ **BASIC ELETRONIC TEST PROCEDURES—2nd Edition—Gottlieb**

The classic test procedures handbook, revised and expanded to include all the latest digital testing and logic probe devices! It covers the full range of tests and measurements. Clearly spelled out techniques are backed up by explanations of appropriate principles and theories and actual test examples. Plus there are over 200 detailed show-how illustrations and schematic diagrams. 368 pp., 234 illus. 7″ × 10″.

Paper $16.95   Book No. 1927
Hard $23.95

☐ **THE FIBEROPTICS AND LASER HANDBOOK**

New video disk players . . . fiberoptic sensing devices used in automobiles . . . these and a host of other new applications for lasers, fiberoptics, and electro-magnetic radiations are presented in this fascinating new sourcebook! Plus, you'll find information on how you can conduct some amazingly simple experiments of your own using both lasers and optical fibers. 368 pp., 98 illus.

Paper $16.95   Book No. 1671

☐ **THE ENCYCLOPEDIA OF ELECTRONIC CIRCUITS—Graf**

Here's the electronics hobbyist's and technician's dream treasury of analog and digital circuits—nearly 100 circuit categories . . . over 1,200 individual circuits designed for long-lasting applications potential. Adding even more to the value of this resource is the exhaustively thorough index which gives you instant access to exactly the circuits you need each and every time! 768 pp., 1,762 illus. 7″ × 10″.

Paper $29.95
Book No. 1938

☐ **THE ILLUSTRATED DICTIONARY OF ELECTRONICS—3rd Edition—Turner & Gibilisco**

The single, most important reference available for electronics hobbyists, students, *and* professionals! Identifies and defines over 27,000 vital electronics terms—more than any other electronics reference published! More than *2,000 new topics* have been added to this state-of-the-art 3rd Edition! *Every* term has been revised and updated to reflect the most current trends, technologies, and usage—with every meaning given for every term! Covers basic electronics, electricity, communications, computers, and emerging technologies! Includes nearly 400 essential drawings, diagrams, tables, and charts! It's the only electronics dictionary that accurately and completely identifies the hundreds of abbreviations and acronyms that have become "standard" in the electronics and computer industries! 608 pp., 395 illus. 7″ × 10″.

Paper $21.95
Hard $34.95
Book No. 1866

*Prices subject to change without notice.

Look for these and other TAB BOOKS at your local bookstore.

TAB BOOKS Inc.
P.O. Box 40
Blue Ridge Summit, PA 17214

Send for FREE TAB Catalog describing over 900 current titles in print.